THE NATURAL GAS SHORTAGE AND THE CONGRESS

Patricia E. Starratt

THE AEI NATIONAL ENERGY PROJECT

The American Enterprise Institute's
National Energy Project was established in early 1974
to examine the broad array of issues
affecting U.S. energy demands and supplies.
The project will commission research into all important
ramifications of the energy problem—economic
and political, domestic and international, private
and public—and will present the results
in studies such as this one.
In addition it will sponsor symposia, debates, conferences,
and workshops, some of which will be televised.

The project is chaired by Melvin R. Laird,
former congressman, secretary of defense,
and domestic counsellor to the President,
and now senior counsellor of *Reader's Digest*.
An advisory council, representing a wide range of
energy-related viewpoints, has been appointed.
The project director is Professor Edward J. Mitchell
of the University of Michigan.

THE NATURAL GAS SHORTAGE AND THE CONGRESS

Patricia E. Starratt

American Enterprise Institute for Public Policy Research
Washington, D. C.

Patricia E. Starratt is a special assistant for legislation in the Federal Energy Administration.

ISBN 0-8447-3148-X

National Energy Study 5, December 1974

Library of Congress Catalog Card No. 74-29378

Printed in the United States of America

CONTENTS

1

THE DEBATE OVER FIELD PRICE REGULATION

The Natural Gas Act, passed in 1938, gave the Federal Power Commission (FPC) authority to regulate sales made by monopoly transmission lines to monopoly retail distribution systems (the "sale in interstate commerce of . . . gas for resale"). Congress did not give the FPC authority to regulate the price paid by pipelines to producers of natural gas (the "wellhead" or "field" price). The act stated that "the provisions of this chapter . . . shall not apply . . . to the production or gathering of natural gas." (15 U.S.C. 717(b)).

The policy was appropriate because there was no "natural monopoly" in need of regulation in the field. Thousands of producers competed for the business of a few major pipelines. Since there were many producers and generally only one pipeline in any one field, the pipeline had market power sufficient to force producers to accept lower prices than would prevail under more competitive conditions.[1]

There might have been an argument for the FPC's setting "floors" under wellhead prices to protect producers, but there was no reason for "ceilings" to be set on such prices. No pipeline operator had any economic motive to pay more than the lowest price for which he could purchase natural gas. Many twenty-year contracts were signed in the 1930s and 1940s at prices as low as two, three, and four cents per thousand cubic feet (Mcf).

The legal authority to regulate field prices for natural gas sold to interstate pipelines emanated from a Supreme Court opinion issued in 1954. Field market competition between pipelines and producers

---

[1] Paul W. MacAvoy, *Price Formation in Natural Gas Fields: A Study of Competition, Monopsony, and Regulation* (New Haven: Yale University Press, 1962), p. 9.

was not, however, the issue involved in the Supreme Court's decision. The decision merely reflected a divided interpretation of the language of the statute.

In the early 1960s regulation began to be effective in that the FPC used its authority to impose ceilings on the field price of natural gas; yet at the same time the commission had a congressional mandate to ensure adequate supply. Real prices for natural gas fell throughout the 1960s, despite the fact that the circumstances affecting natural gas development had changed in three important ways. First, as pipelines expanded in numbers and distance, a national market for natural gas had developed. The demand for natural gas doubled between 1957 and 1972. Second, producers were developing an ability to separate their search for natural gas from their search for oil: natural gas was no longer simply a by-product of oil exploration and production. In fact, by 1972, 78.9 percent of the natural gas produced came from gas wells. Third, as time passed, the amount of reserves found per foot drilled predictably declined as the easiest, largest prospects are, where possible, drilled first. Higher natural gas prices were thus increasingly needed to elicit new natural gas supplies.

Overview of the Twenty-year Debate

The two goals of low prices and adequate supply are not necessarily compatible in any business. Low prices for a valuable commodity will (*ceteris paribus*) raise the quantity demanded and lower the quantity supplied. In the natural gas production industry, low prices, as costs rise, increasingly restrict the number of projects it is profitable to explore and develop and decrease the rate of replacement of depleted fields. The compatibility of the two goals has been one of the main themes of the debate that began with the 1954 Supreme Court decision.

For two decades, a battle has raged at the FPC, in the courts, and in the academic world over the field price regulation of natural gas sold interstate. Several times, Congress has become involved in the debates. But some of the issues discussed at the congressional level have in fact been resolved at the regulatory level.

While the importance of congressional action in 1974 is obvious, political discussions of natural gas policy are not always especially useful. An artificial distinction is usually established: one must be either "pro-industry" or "pro-consumer." But the suggestion that there is an "industry view" is inaccurate.

Industry Groups. For one thing, there are various industry groups (producers, pipelines, and distributors) with differing and even opposing interests. It is well known that producers have an interest in higher prices for natural gas (gas) sold at the well. It is less well known that most nonproducing gas utilities (pipelines and distributors) have a strong economic incentive to keep field prices down.[2]

In order to increase revenue and profit, a public utility must increase its rate base,[3] because the rate base determines the earnings the utility is allowed. A utility does not increase its rate base by passing on higher-cost gas to consumers. Low-priced gas encourages growth in demand. Without growth in sales and in capital facilities, the rate base depreciates. With asset growth, a utility can gain growth in revenues without incurring adverse regulatory action.

The need for growth in the rate base explains in large measure the utilities' past effort to hold prices paid to producers below competitive levels. As Clark Hawkins has noted,

> the existence of the possibility of higher growth and higher profit here is important because it provides an explanation of the vigorous intervention in field price regulation by the gas distributors.[4]

Disparity among Firms. A second factor which makes any notion of an "industry view" inaccurate is the disparity among the firms in the industry. Despite the appearance of solidarity in utility presentations in congressional hearings on proposed natural gas legislation, individual pipelines and distributors differ as to what legislation should be enacted to deal with the natural gas shortage.

Seventeen interstate pipelines, unable to obtain enough gas to meet their firm contract requirements, were forced to curtail deliveries in 1972 and 1973. Net curtailments totaled 1,191,132 Mcf for the year April 1973 through March 1974, equivalent, on a Btu basis, to the energy loss from the oil embargo during the first quarter of 1974. Net supply deficiencies totaling 1,845,770 Mcf have been projected for the year from April 1974 through March 1975, exceeding the preceding year's curtailments by 654,638 Mcf.

---

[2] Robert Helms, *Natural Gas Regulation: An Evaluation of FPC Price Controls* (Washington, D. C.: American Enterprise Institute for Public Policy Research, 1974).

[3] This assumes that the allowed rate of return at least covers their cost of capital. If it does not, additional investment will cause losses.

[4] Clark A. Hawkins, *The Field Price Regulation of Natural Gas* (Tallahassee: Florida State University Press, 1969), p. 197.

Several pipelines are in very deep curtailment. Three major pipelines (Arkansas Louisiana Gas Co., Trunkline Gas Co., and United Gas Pipe Line Co.) each curtailed approximately one-third of their firm contracts in the 1972 and 1973 period. Four other companies curtailed between 15 and 20 percent of their contracts. The 1974–75 winter outlook for United Gas Pipe Line Company (United) is particularly severe. Under normal winter conditions, United may have to curtail volumes as much as 41 percent. Much of United's gas is sold to seven other interstate pipelines. Indirectly, therefore, United's cutbacks will affect consumers in some 20 states.

Curtailments of this sort are predicted to lead to cutbacks in deliveries to third-priority customers (such as, for example, small industrial users with firm contracts) in the summer of 1975. In some areas, deliveries to second and first priority customers (commercial and residential) could well be curtailed within the next year.

Although the demand during the winter of 1973–74 was not unusually high, curtailments increased by 39.46 percent over 1972 and 1973, with Arkansas Louisiana, Trunkline, and United Gas suffering the worst deficiencies. As a consequence of inability to meet contract commitments, most pipeline companies, formerly opposed to any increase in the field price of gas, have not only withdrawn their opposition but favor phased deregulation of natural gas field prices.

Some utilities are committed to projects for importing liquefied natural gas (LNG) or building synthetic natural gas (SNG) plants.[5] These capital-intensive projects will increase not only the utilities' gas supplies but also could increase their rate bases. From an economic standpoint, companies with projects of this sort are less interested in increasing conventional domestic gas supplies through higher field prices than are utilities that have become involved in the natural gas production business themselves,[6] and companies that are seeking to purchase the natural gas required to serve their present and prospective customers.

---

[5] Plans for at least seventy-one LNG and SNG projects were on file at the FPC in 1973. As of June 1974, there are at least thirty-four SNG projects which are operational, under construction, or in various stages of planning. Whether or not some of these plants will obtain desired quantities of feedstocks is open to question because of the Mandatory Allocation Act administered by the Federal Energy Administration. As to the five LNG projects currently pending before the FPC, the price to be paid for the LNG imports is open to question. Previously contracted prices are either under renegotiation or are expected to require revision in the future as a condition to approval by the exporting countries involved (Algeria and Indonesia).

[6] U.S. Senate, Committee on Interior and Insular Affairs, *Natural Gas Policy Issues and Options*, Serial No. 93-20 (92-55), 1973, pp. 81-82.

Deregulation and the Consumer. A third reason why it is inaccurate to discuss possible courses of action as "pro-industry" or "pro-consumer" is that recognized experts (from all shades of the political spectrum, including more than a few who actively supported Senator George McGovern in his 1972 presidential campaign) have determined that the deregulation of field prices will benefit consumers far more than regulation which is touted as consumer-oriented. (See Table 1.) But since actual consumers (as opposed to consumer advocates) are rarely heard in congressional debates, it may be worthwhile identifying natural gas consumers and their interests to better analyze policy options.

Generally, mention of "natural gas consumers" calls to mind Mrs. Jones cooking on her gas stove and paying the bills for her gas service, but the bulk of natural gas consumed in the United States is not used in homes. The 1972 Bureau of Mines Mineral Industry Survey showed that industry used 41.6 percent of all gas consumed; electric generating facilities used 20.3 percent; residential consumers used 25.3 percent; and commercial establishments used 11.1 percent. While the number of consumers in one category or another differs regionally, it is noteworthy that bulk industrial and electric utility users together accounted for 61.9 percent of natural gas consumption. Although natural gas accounted for 32.3 percent of domestic energy used in 1972 and was delivered to more than 43 million consumers, supplies to current residential users are generally not threatened by shortages at this time. Potential residential users have, however, been unable to obtain gas service in Washington, D. C., for example; this situation has also existed for several years in other areas of the country. The waiting list in Chicago in 1971 included 14,000 residences; 2,000 businesses; 800 industrial consumers.[7]

Supply reductions to current customers are increasing quickly, both in volume and area. From April 1972 to March 1973, 825 billion cubic feet of gas were curtailed by major interstate pipeline companies. From April 1973 to March 1974, estimated curtailed volumes increased to 1,200 billion cubic feet (Bcf), equivalent to 6 percent of total natural gas production or 200 million barrels of oil. Net supply deficiencies for November 1974 through March 1975 are estimated at 919 Bcf, compared with actual net curtailments for the 1973 and 1974 heating season of 424,520 Mcf, an increase of almost 110 percent. As of October 1974, virtually every state in the nation is affected, especially in Region Three. (See Table 2.)

---

[7] Patricia E. Starratt, "We're Running Out of Gas—Needlessly," *Reader's Digest*, April 1973, pp. 167-171.

Table 1

SUMMARY OF CONSUMPTION, WHOLESALE AND RESIDENTIAL PRICES (1972 DOLLARS) UNDER REGULATION AND DEREGULATION OF WELLHEAD PRICES

	Units	Regions									
		1	2	3	4	5	6	7	8	9	10
Wholesale Price (1972$)											
1971	$/Mcf	.71	.46	.39	.39	.43	.39	.27	.29	.38	.35
1980, deregulation	$/Mcf	.90	.70	.60	.63	.70	.60	.63	.54	.61	.58
1980, regulation	$/Mcf	1.05	.90	.87	.77	.79	.48	.54	.42	.58	.70
Consumption											
1980, deregulation	Tcf	.304	4.527	2.034	3.965	.957	1.785	7.444	.532	2.590	.433
1980, regulation	Tcf	.247	4.114	1.826	3.515	.936	2.377	9.510	.631	2.641	.342
Shortfall											
1980, regulation	Tcf	.158	1.923	.945	1.774	.340	.003	.417	.078	.813	.235
	% of dem.	39%	32%	34%	34%	27%	—	4%	11%	24%	41%
Residential Price (1972$)											
1971	$/Mcf	1.57	1.591	.986	1.22	.82	1.05	.54	.65	1.33	1.50
1980, deregulation	$/Mcf	1.75	1.83	1.196	1.46	1.09	1.26	.90	.90	1.56	1.73
1980, regulation	$/Mcf	1.91	2.03	1.468	1.60	1.18	1.14	.81	.78	1.53	1.85
Residential Price Increase Due to Deregulation	%	−8%	−10%	−18%	−9%	−8%	11%	11%	15%	2%	−6%

Source: *Final Environmental Impact Statement: Proposed Deregulation of Natural Gas Prices* (Washington, D. C.: U.S. Department of the Interior), June 1974, p. G-2.

Table 2

SUMMARY OF CURTAILMENTS FOR THE FIVE-MONTH PERIOD EXPRESSED AS TOTAL VOLUMES (BCF) AND PERCENTAGES OF TOTAL REQUIREMENTS BY FPC REGIONS

(November 1974 through March 1975)

	Region	Volumes (Bcf)	Percentages
1	Connecticut, Maine, Massachusetts, New Hampshire, and Rhode Island	6.6	4.2
2	Delaware, District of Columbia, Kentucky, Maryland, New Jersey, Ohio, Pennsylvania, New York, Virginia, and West Virginia	184.2	8.5
3	Alabama, Florida, Georgia, North Carolina, South Carolina, and Tennessee	148.8	18.1
4	Illinois, Indiana, Michigan, and Wisconsin	91.0	6.1
5	Iowa, Minnesota, Nebraska, North Dakota, and South Dakota	1.9	1.3
6	Kansas, Missouri, and Oklahoma	78.1	17.4
7	Arkansas, Louisiana, Mississippi, and Texas	177.9	44.6
8	Colorado, Montana, Utah, and Wyoming	29.8	15.1
9	Arizona, California, Nevada, and New Mexico	27.6	9.9
10	Idaho, Oregon, and Washington	46.3	25.0
		792.2	12.6

Source: Data received as of 23 October 1974, in response to Federal Power Commission's letter requesting estimated requirements, curtailments and deliveries to customers of 42 interstate pipeline companies for the period November 1974-March 1975.

The Btus lost as a result of the gas curtailment from April 1973 to March 1974 (1,191 billion cubic feet), are equivalent to the maximum impact of the oil embargo during the first quarter of 1974 or about 2 million barrels per day, as reported by the National Petroleum Council.

Natural gas curtailments present serious problems for the industrial consumers affected. In the Baltimore area, a natural gas curtailment order would directly result in the loss of 10,000 jobs, according to a recent study.[8] New Jersey also predicts severe economic impacts from the gas shortage. If curtailments increase from 28 per-

[8] "Joint Petition of Maryland Group of Industrial Consumers of Natural Gas for Leave to Intervene," Docket No. RP72-89, Federal Power Commission, 22 April 1974, p. 2. Mimeographed.

cent to 30 percent, the state predicts that 120,000 workers might be laid off. Indirectly, another 40,000 workers could be affected.[9] Natural gas curtailments do have a ripple effect, contributing to unemployment, reducing national output and raising prices for other commodities.

It is especially ironic that, at a time when the nation needs more steel for energy-related activity, steel companies are facing production cutbacks for lack of natural gas. The Standard Steel Company in Pennsylvania, for example, is the sole supplier of another company which manufactures equipment for nuclear power plants and also plays a vital role in the production of steel for the transportation of coal. On 21 October 1974, the company's gas supplies were curtailed—with only a few days notice—100 percent. The immediate layoff affected 600 employees. Through urgent requests for relief, the crisis situation was temporarily resolved, in part because Standard Steel could convert 35 percent of its operation to Number 4 fuel oil.

Many companies, however, lack the capability to switch to other fuels. Some smaller consumers cannot afford the high cost of conversion. There is also the serious problem of inadequate storage capacity for alternate fuels as well as the problem of obtaining alternate fuels that can be used. In 1973, the United States produced less coal than it did in 1947. Domestic oil production has been declining. State air quality standards limit the use of alternate fuels. According to studies of the National Academy of Sciences and the Environmental Protection Agency, the national capability to install sulphur fuel is limited to a maximum of fifty-five installations over the next five years. There are no easy answers to the supply problems facing natural gas consumers in the United States.

Indirectly, shortages of natural gas affect the supply and price of other fuels for all consumers. With the demand for low sulphur oil and coal increasing because of natural gas shortages, prices for the unregulated fuels doubled between 1970 and 1972. In 1972 and 1973 natural gas utilities and large industrial users increased their purchases of propane (a natural gas liquid) as a replacement for curtailed gas supplies, and paid higher prices to get it. This led to propane shortages, and resultant industrial stockpiling kept farmers from buying the propane they needed to harvest and dry their crops.

One way or another, all consumers will pay more indirectly or directly for energy, even if the choice is to make do with less energy.

---

[9] Henry A. Watson, Assistant Director, Office of Business Economics of the New Jersey Department of Labor and Industry, in a 22 July 1974 memorandum to Joseph Coviello, Director, New Jersey Department of Labor and Industry, p. 3.

Consumers will ultimately pay no more or less than what they estimate to be the value of the product they seek when that product is in short supply. Whereas in past years, unregulated intrastate buyers paid no more than interstate buyers in the field,[10] intrastate prices have gone up in recent years as a result of the scarcity of natural gas and the high cost of alternative fuels. If industrial customers are lost (either by curtailment or by loading higher-priced "new" gas costs solely on industrial customers), the pipelines' increased costs from operating at less than full capacity will be paid for by residential consumers.

Even if natural gas imports and substitutes could fill the gap between conventional gas and current demand, prices would still rise. Indeed, the increase in the cost of gas as a consequence of the use of SNG and LNG will, by 1980, cause consumers in most regions to pay higher prices than they would with deregulation, according to a study by the U.S. Department of the Interior.[11] In Region One (the Northeastern United States), for example, the residential price of gas would be 15 cents per Mcf higher (in 1972 dollars), with regulation, assuming SNG and LNG cost no more than expected in early 1973, than it would be without regulation. (See Table 1.) And if the higher prices for gas do not fall equally on all users of natural gas, industrial and electric utility users will increase the price of goods and services to all consumers who use electricity, drive automobiles, buy pharmaceutical products, and so on. From the point of view of the informed consumer, the relevant issue is not, therefore, "Will Congress protect me from paying higher prices?" but rather "What are the costs and benefits of the major natural gas policy options?"

The Debate to 1974. The debate over field price regulation was summarized by Clark Hawkins in 1969:

> Those pro-regulation argue that a small number of producers make the bulk of the sales and control most of the reserves. Thus monopoly power is present. Since natural gas lines are very expensive and permanent, they can't be moved around like "garden hose." Pipelines bid against each other for reserves and the price spirals up. The consumer is not protected from exploitation. Those anti-regulation argue

---

[10] Robert W. Gerwig, "Natural Gas Production: A Study of the Costs of Regulation," *Journal of Law & Economics*, vol. 5 (October 1962), pp. 69-92.

[11] U.S. Department of the Interior, *Final Environmental Impact Statement: Proposed Deregulation of Natural Gas Prices*, Office of Economic Analysis, June 1974. See table p. G-2.

that the field markets are competitive, and costs are not being recovered. The consumer will pay higher prices in the long run with regulation than without. Only 10% of the burner tip price of gas is for the gas itself; 90% is for service, transportation, distribution. Utilities fight raises in the field, not out of interest in the consumer, but because a rise in gas costs hampers their expansion of sales, and impairs their competitive position with coal, oil, electric.[12]

In the words of Alfred Kahn in 1960, the debate on each side is based on the following logic:

> *Opponents of Regulation:* Competition is effective; and that is all one needs to know; even if it were not, supply and demand are elastic; even if they were not, it makes no sense to regulate on a cost of service basis.
>
> *Proponents of Regulation:* Competition is not effective; even if it were, supply and demand are inelastic; even if they were not, it is desirable to regulate the price of gas on a cost of service basis to restrict profits and prevent exploitation by obtaining gas for the consumer at the lowest possible price.[13]

The major issues have historically concerned

(1) The degree of competition among producers of natural gas in the field;

(2) The elasticities of supply and demand;

(3) The economic usefulness of trying to ascertain the cost of producing natural gas;

(4) The efficacy of field price regulation: keeping prices low for the consumer; the effect on supply.

The Congressional Debate, 1973 and 1974

The focus of the debate and the issues changed somewhat in 1973 because of the growing imbalance between natural gas supply and demand. While the question of competition was heavily stressed, a key issue was the nature of the shortage: was it real or contrived? If the shortage was real, it would seem that regulation must be the primary cause.

---

12 Hawkins, *The Field Price Regulation of Natural Gas*, p. 143.

13 Alfred E. Kahn, "Economic Issues in Regulating the Field Price of Natural Gas," *American Economic Review*, vol. 50. no. 2 (May 1960), pp. 506-525, as paraphrased in Hawkins, *The Field Price Regulation of Natural Gas*, pp. 143-144.

Those who favored continued regulation sought out witnesses to give evidence that it was contrived "by the monopolistic oil companies." Where they did not wish to use the word "collusion," they spoke of the "refusal" of the oil companies to develop "known reserves." As shortages of fuel oil had developed in 1973 before the Arab embargo, natural gas shortages were explained by reference to the marketing practices of the oil companies. The implication was that companies misallocating short supplies to independent oil jobbers must be corrupt—and able to get away with it—in their field market behavior. It was pointed out that there were shortages of other fuels, and that these fuels were not regulated by the FPC. (The logical conclusion was said to be that natural gas shortages therefore cannot be related to regulation, a conclusion that ignores the less obvious means through which real petroleum prices were also frozen for a decade.)

Two key witnesses from the FPC's Office of Economics, David S. Schwartz and John W. Wilson, testified that (1) the costs of producing natural gas can be ascertained with relative ease with the appropriate data and (2) the structure of the oil industry is such that there is virtually no competition in field markets. A final argument for regulation suggested that, with deregulation, wellhead prices for natural gas would rise to levels comparable to the crude oil prices set by the producing-nation cartel, the Organization of Petroleum Exporting Countries (OPEC).

Few of the academic experts who had predicted natural gas shortages years before as the inexorable outcome of field price regulation were called to testify in congressional hearings. The testimony of those who did appear (notably Edward W. Erickson and Paul W. MacAvoy) was devoid of emotion-laden phrases such as "the multinational oil companies," and "windfall profits at the expense of consumers." But, though heard by only a few, their testimony made good economic sense.

The ultimate goal of some of those who argued in favor of regulation turned out to be the eventual nationalization of the oil and gas industries. This would be accomplished through (1) the creation of a federal drilling corporation that (with its advantages over private enterprise) would eventually dominate domestic petroleum exploration and development, and (2) the extension of FPC jurisdiction to the unregulated intrastate market, both to control field prices nationwide and to allocate natural gas away from industrial users in the producing states to residential users in the areas where shortages are expected to become more acute over time.

What Is at Stake for Consumers and Legislators in 1974?

The Arab oil embargo made the 1974 natural gas policy debate critical. Those who have wanted to blame the oil industry for the natural gas shortage now see an opportunity to blame the oil industry for the entire "energy crisis." The public at large is confused about the nature of the energy shortages and their "cause." They are looking for an easy answer. Yet any easy answer to the nation's multifaceted energy dilemma is likely to be inaccurate.

With inappropriate policies, supplies of domestic natural gas will fail more and more to meet demand. Government and industry forecasts in the last few years suggest that by 1980, the nation will be dependent on foreign oil for 50 percent or more of its energy supply. Demand projections are, of course, subject to various assumptions, but it is worth remembering that in the past, both industry and government underestimated the growth of energy demand, particularly the growth of the demand for natural gas.

In 1973, 22.8 trillion cubic feet (Tcf) of natural gas were consumed in the United States. FPC Staff Report No. 2 estimates that the demand for gas will exceed supply by 9.5 Tcf in 1980, 13.7 Tcf in 1985, and 17.1 Tcf in 1990. The estimates take into account 1972 projected gas supplies from supplemental sources, increased levels of successful exploration, and more moderate growth rates than those occurring in previous years. To balance the anticipated deficit, average annual reserve additions would have to be half again the 1956 all-time record of reserve additions from the lower 48 states (24.7 Tcf).[14]

The gap between supply and demand is likely to be more severe because of the repercussions of the international political events of 1973. The volumes of liquefied natural gas (LNG) that were expected to be delivered between 1971 and 1985 were (1) 29.4 Tcf, North America; (2) 5.0 Tcf, other Western Hemisphere; (3) 25.06 Tcf, Eastern Hemisphere. The largest volume to be imported from a single country was 8.56 Tcf from Algeria. Second in projected volume was 6.57 Tcf from the U.S.S.R.

Synthetic natural gas derived from liquid hydrocarbons (twenty-eight projects were in various stages of development in 1972) depends on the cost and availability of feedstocks. Most projects require naphtha, a crude oil derivative, for feedstock. The prospect for dependable naphtha supplies has diminished sharply in the wake of the oil embargo.

---

[14] Federal Power Commission, Bureau of Natural Gas, "Natural Gas Supply and Demand, 1971 to 1990," Staff Report No. 2, 1972.

In time, other sources of energy will be available as dependable fuels. FPC Chairman John N. Nassikas has predicted that by 1990, nuclear power will supply 23 percent of the national energy requirements (equivalent to 31 Tcf). Past Chairman Joseph C. Swidler of the New York Public Service Commission may be correct, however, in his statement that this prediction is "grossly exaggerated." Despite twenty years of research and $800 million in government funding over a four-year period, nuclear power supplies only 2 percent of the energy used in the United States. Long-term answers to the supply question may well include geothermal energy, solar energy, oil shale, tar sands, and nuclear energy. In the short term, conservation is imperative. But even so, the economic life of the United States for the decade ahead depends to a substantial degree on the passage of appropriate natural gas legislation.

2

THE NATURAL GAS SHORTAGE

Contrived?

In hearings held pursuant to Senate Resolution 45 authorizing a National Fuels and Energy Policy Study, the Interior and Insular Affairs Committee heard testimony from a number of witnesses as to the reliability of estimates of "proved reserves." Bruce C. Netschert (National Economic Research Associates, Inc.) concluded that there is no evidence to substantiate the view that reserve figures were deliberately manipulated.[1] King Hubbert, formerly with the United States Geological Survey (USGS), maintained that proved reserve estimates are "probably about as accurate as engineering knowledge will permit."

> With regard to deliberate falsification, it must be borne in mind that those compilations are made by committees of engineers employed by different and competing oil and gas companies. Hence, were any company to attempt to falsify its data, this would almost immediately be recognized by the engineers on the same committee of rival companies operating in the same field and areas. . . . It is unavoidable that an estimate of this kind must have a considerable latitude of uncertainty and error. However, by repeating the estimates at 2-year intervals, it is expected that initial errors will be corrected and that the final estimates will converge to the true ultimate production. . . . In summary, it is my view that the natural gas statistics, where based on measurable quantities, are about as accurate as can be expected.[2]

[1] Bruce C. Netschert, "Explanation of Negative Reserve Additions," *Natural Gas Supply Study,* Senate Interior and Insular Affairs Committee (37-792), 1972, pp. 91-94.

[2] M. King Hubbert, enclosure to letter of Hollis M. Dole, written in response to

Testimony of a gas distributor before the Senate Commerce Committee corroborates the view that the gas shortage is indeed real:

> My own company, the Consolidated Natural Gas System, engaged the three largest geological consulting firms in the country to tell us if gas were being held back. The answer was no. Undoubtedly, numerous other gas companies did the same as well as utilized their own staff of engineers, geologists, and petroleum experts to ascertain that there were no unknown significant proved reserves. And they can look to data that must be filed with the states for all wells drilled in performing these checks.
>
> Nor is this negative answer surprising. There are numerous pressures on a producer who finds and develops a gas reserve to market it immediately. One is to receive current income as soon as possible and enhance the present value of the property on which great sums have been spent over a period of probably from five to seven years.
>
> Another is the lessor's obligation to the landowner pursuant to the lease agreement. Once a capability of production is established there must be a prudent effort to market the production to perpetuate the lease. Since the landowner looks to royalties from production, this is an obligation that he, his consulting engineer or geologist, and his lawyer are very aware of. Also, if production is initiated by an offset lease owner who has the same pressures on him, the failure to promptly do likewise results in drainage of valuable reserves. I do not profess to be an expert in the field of petroleum leasing, exploration and development. However, as an executive who looks to these experts for my vital supplies, I am convinced it is difficult, if not impossible, to find and develop any significant gas reserves in secrecy. Therefore, my own checks and analyses convince me as a gas buyer and distributor to our consumers that the reported reserve data, and hence the shortage, is quite valid.[3]

The FPC has also conducted field studies of gas reserve estimates. The most significant in scope and conclusions was the National Gas Survey. The survey concluded that the industry's proved reserve estimates (as collected and compiled by the American Gas Association) were 9 percent too high in 1970.

---

questions of Senator George McGovern, Senate Interior and Insular Affairs Committee, 15 December 1969, reprinted in *Natural Gas Supply Study*, Senate Interior and Insular Affairs Committee (37-792), 1972, pp. 40-41.

[3] G. J. Tankersley, statement of the American Gas Association before the U.S. Senate Commerce Committee, 8 November 1973, pp. 12-13. Mimeographed.

Despite the use of an independent accounting agency and governmental and academic staff support to conduct the survey, some have questioned the validity of its findings. Those who believe the survey's results cannot be trusted refer to testimony of Dr. Howard William Pifer III (a member of the Technical Advisory Committee for Supply and the Statistical Validation Team). In his 26 June 1973 statement before the Subcommittee on Antitrust and Monopoly, Dr. Pifer raised doubts as to a number of aspects affecting the field study of reserves.

The reliability of Pifer's testimony was called into question by Frank C. Allen, chief of the FPC's Bureau of Natural Gas as follows:

> Dr. Pifer, a former faculty member of Harvard University, was acting as a statistical consultant under contract to the FPC. Dr. Pifer's participation in the National Gas Reserve Study was limited to the activities of the team which was to prescribe sampling procedures for valid reserve estimation. His services were not obtained on the basis that he was knowledgeable on the estimation of gas reserves and there is nothing in his credentials to indicate that he is competent in this area. Dr. Pifer has testified relative to functions of the National Gas Reserve Study that are beyond his area of expertise.[4]

In an extensive analysis of the testimony given by Pifer and James T. Halverson, director of the Bureau of Competition at the Federal Trade Commission, Dr. Paul J. Root, technical director of the National Gas Survey, pointed out areas in which what was said was inaccurate, incorrect, or misleading. The following are a few examples:

> Mr. Halverson makes the statement that "it is our experience that the raw data necessary in order to make original estimates of natural gas reserves cannot be obtained outside of company sources and that when the Federal Power Commission relies solely on the company (without independent audit) to provide all the raw data, that reliance is misplaced." I wish to emphasize again that the FPC field reserves teams did, in fact, make an independent estimation of the reserves for all fields comprising the statistical sample. The field team leaders are either trained geologists or petroleum engineers who have had extensive experience in making reserve estimates as part of their duties with the Federal Power Commission. . . .

[4] Frank C. Allen, comments on "Policy Issues and Options Affecting Natural Gas," in *Natural Gas Policy Issues and Options,* Senate Interior and Insular Affairs Committee, 1973, p. 157.

> As noted by Mr. Halverson in his criticism of the reporting procedures used by the American Gas Association (AGA), "Even this review [of the reserves estimates compiled for the fields] is limited in the sense that Subcommittee members only see the final estimate for a particular field and they do not see the underlying data." To overcome this very deficiency the NGRS [National Gas Reserve Study] was designed so that the field reserves teams could see *and analyze* the underlying data. Perhaps an equally significant implication of Mr. Halverson's criticism is that it applies with the same force to the procedures apparently being used by the staff of the Federal Trade Commission. . . . However, a more serious consideration and perhaps even a critical infirmity in the FTC investigation is that the judgmental decisions made by industry personnel who prepared the reserves estimates have been untested or unchallenged. The FTC investigation might be characterized as one involving the processing of "paper reserves," whereas the NGRS is based on independent estimates of proven reserves actually in the ground at a certain point in time.[5]

It appears that Halverson's chief problem has been that of comparing apples and oranges when comparing reserve data or, perhaps more accurately, seeds, green apples and red apples. The National Gas Survey and the American Gas Association have specific definitions of "proved reserves." The definition of the AGA's Committee on Natural Gas Reserves is as follows:

> The current estimated quantity of natural gas liquids, which analysis of geologic and engineering data demonstrate with reasonable certainty to be recoverable in the future from oil and gas reserves *under existing economic and operating conditions*. Reservoirs are considered proved that have demonstrated the ability to produce by either actual production or conclusive formation tests. [Emphasis added.] [6]

Proved reserves, in other words, are those quantities of gas known to exist with the highest degree of certainty. They do not include quantities which may with reasonable certainty be known to exist but which do not come within the AGA's definition. (The definition of "proved reserves" used by AGA does not include, therefore,

---

[5] Paul J. Root, letter to Senator Philip A. Hart, 27 July 1973, pp. 8-9.

[6] *Reserves of Crude Oil, Natural Gas Liquids and Natural Gas in the United States and Canada: Productive Capacity as of December 31, 1970*, vol. 25 (May 1971), p. 102. This is a joint publication of the American Gas Association (AGA), the American Petroleum Institute (API), and the Canadian Petroleum Association.

"potential" or "speculative" reserves that are suggested by geologic tests performed before drilling and before bidding on new acreage in federal offshore lease sales.) Most important, they do not include those resources which are known with reasonable certainty to be recoverable, but only under changed "economic conditions," that is to say, higher prices. The following assumptions were used in the National Gas Survey:

> If sold in interstate commerce, sales prices for gas will be at the effective rate as of December 31, 1970, (or at FPC ceiling if the gas is not under contract) with no allowance for price escalations beyond those already approved in the FPC area rate orders. . . .[7]

Reserves which could economically be produced at 45 cents per Mcf in Southern Louisiana, but not at the 26 cents per Mcf ceiling price which had been set for the area, by definition are not "proved reserves."

Halverson's statement that AGA-proved reserves are lower than reserve figures quoted in Form 15 reports filed at the FPC and lower than in-house company estimates shows a similar confusion on reserve data. When a producer is making a decision whether to bid on a lease, or build a drilling platform, or even sell reserves to a pipeline, the producer's reserve data may include reserves classified as "probable." There have been a number of FPC cases involving pipelines that contracted with producers in fields where only a few exploratory wells had been drilled. The reserves underlying a supply contract with a pipeline were still classified as "probable" because of the lack of development of the field. Company and Form 15 reserve figures often differ substantially from AGA "proved reserve" figures because they do not represent the same reserves. Form 15 data represent "salable" reserves. Individual company definitions vary from company to company and differ both from AGA definitions and Form 15 definitions.

In sum, the evidence used to support the view that the natural gas shortage is contrived does not stand up under analysis. Use of FPC Form 15 data, company data, and AGA estimates in determining the reliability of reserve estimates is simply wrong, since the definition of "proved reserves" is applicable only to the AGA figures. In the FPC's National Gas Survey, AGA estimates were only used to define the statistical population so that a frequency distribution could be developed for sampling purposes. In the words of Root, "it is true

[7] Root, letter to Senator Hart, p. 6.

that distortions in the population would be manifest in the frequency distribution, and that they would affect the reliability of the estimate; however, they would have relatively little effect on the projected total reserves." [8]

It is also noteworthy that the total uncommitted reserves of the sixty-nine producers who reported to the FPC as of 30 June 1973, pursuant to FPC Docket No. R-405, amounted to only 1.15 percent of the total proved reserves for the lower forty-eight states. These uncommitted reserves are equal to about one-fourth of the recent annual gross additions to reserves. The average time elapsing between availability of additional proved reserves and commitment to contract is "about three months," [9] the time required to consummate a sale.

While major producers conceivably might try to hold back discovered gas (and may indeed be doing so to some degree), it is unlikely that there is enough gas available but not being sold to account for the shortage being experienced by consumers even in the unregulated intrastate market. It is highly unlikely that the major companies possess sufficient market power to control the output of the top 101 producers (each with annual sales of 10 billion cubic feet or more), the 1,095 middle-size producers (each with annual sales of 2 billion cubic feet but less than 10 billion cubic feet), and the 3,449 small producers (each of whom sold less than 2 billion cubic feet).[10]

Real?

There are significant indications (not depending on the published estimates of proved reserves) that the shortage has not been contrived by producers. Distributors have historically fiercely opposed field price increases and deregulation. Yet they are now convinced that there is in fact a shortage, and they are going to considerable trouble and making long-term commitments to try to line up LNG and SNG supplies.

The liquidation of the natural gas surplus that existed in the 1940s and 1950s, and an eventual excess of demand over supply could, moreover, be predicted as the inexorable outcome of (1) prices

---

[8] Ibid., p. 5.

[9] Celia Star Gody, attachment to letter of W. O. Senter to Senator Philip A. Hart, U.S. Senate Committee on the Judiciary, 6 December 1973, p. 2.

[10] *The Gas Supplies of Interstate Natural Gas Pipeline Companies, 1971* (Washington, D. C.: Federal Power Commission, 1972), p. 102.

for natural gas significantly lower than prices for other fuels; (2) the "topping out" of domestic oil production (and thereby the supply of dissolved and associated natural gas); and (3) the progressive discovery and depletion of the best and least costly onshore reserves and prospects. Even if the current reserve estimates were deliberately shaded to create an illusion of shortages, these three factors guarantee that there will be a real and growing shortage.

If producers were restricting output to force a price rise, one would expect to find production declining relative to reserves, with a consequent rise in the reserves-to-production ratio. For the past six years, the reserves-to-production ratio has been falling. (See Table 3.)

The American Gas Association first began keeping records of changes in natural gas reserves in 1946. Two ratios have been developed to indicate supply availability: (1) the ratio between the amount of new reserve additions ("findings") and the level of production (the F/P ratio) and (2) the ratio between proved reserves and annual production (the R/P ratio).

Between 1946 and 1970, new proved reserve additions increased by 63 percent, while production increased by 344 percent. In the last six years, the amount of natural gas produced has been twice as great as the amount of proved reserve additions to the inventory. If natural gas prices were adequate, investment would be attracted to drilling for additional reserves and the additions to reserves would be sufficient to offset withdrawal. The result would be at least a maintenance of the reserve-to-production ratio rather than a decline in the ratio.

Since 1963, moreover, the FPC has been using pipeline company data to assess the physical capability of supply sources for continued gas delivery. Over time, the production or "deliverability" (future capacity to deliver) of a well declines, either from changes in the physical condition of the well (decreased pressure for example) or because production has been at a faster rate than originally anticipated. The combination of steady increase in market demand (consumption rates doubled between 1957 and 1972) with insufficient new additions to reserves in the 1960s resulted in an accelerated depletion of reserves and falling deliverability rates. In December 1972, the FPC reported that "the composite 1971 estimates indicate that the presently owned and contracted for gas reserves can no longer sustain any increased market requirements. . . .[11]

[11] Ibid.

Table 3

U.S. NATURAL GAS SUPPLY EXCLUDING ALASKA, 1946 AND 1972[a]

(Trillions of cubic feet)

Year (1)	Net Production (2)	Reserve Additions (3)	Year-End Reserves (4)	R/P Ratio (col. 4 divided by col. 2) (5)	F/P Ratio (col. 3 divided by col. 2) (6)
1946	4.9	17.6	159.7	32.6	3.6
1947	5.6	10.9	165.0	29.5	1.9
1948	6.0	13.8	172.9	28.8	2.3
1949	6.2	12.6	179.4	28.9	2.0
1950	6.9	12.0	184.6	26.8	1.7
1951	7.9	16.0	192.8	24.4	2.0
1952	8.6	14.3	198.6	23.1	1.7
1953	9.2	20.3	210.3	22.9	2.2
1954	9.4	9.6	210.6	22.4	1.0
1955	10.1	21.9	222.5	22.0	2.2
1956	10.9	24.7	236.5	21.7	2.3
1957	11.4	20.0	245.2	21.5	1.8
1958	11.4	18.9	252.8	22.2	1.7
1959	12.4	20.6	261.2	21.1	1.7
1960	13.0	13.8	262.2	20.2	1.1
1961	13.4	16.4	265.4	19.8	1.2
1962	13.6	18.8	270.6	19.9	1.4
1963	14.5	18.1	274.5	18.9	1.2
1964	15.3	20.1	279.4	18.3	1.3
1965	16.3	21.2	284.5	17.5	1.3
1966	17.5	19.2	286.4	16.4	1.1
1967	18.4	21.1	289.3	15.7	1.1
1968	19.3	12.0	282.1	14.6	.6
1969	20.6	8.3	269.9	13.1	.4
1970	21.8	11.1	259.6	11.9	.5
1971	21.9	9.4	247.4	11.3	.4
1972	22.4	9.4	234.6	10.5	.4

[a] Data represent total U.S. natural gas supply prior to 1960. Alaska's natural gas supply was not reported until 1960.

Source: American Gas Association, *Reserves of Crude Oil, Natural Gas Liquids and Natural Gas in the United States and Canada*, as of 31 December 1972.

In short, the statistical data submitted by the industry to the FPC suggest that the reserve shortage is the result of declining trends in reserve additions at a time when overall production was going up. Total proved reserves have not only failed to keep up with demand; they have actually declined (Table 3). The unprecedented and unforeseen growth in natural gas demand served to increase the already growing stress on limited reserves.

To the extent that the discovery of natural gas (particularly gas found associated with oil in the well) has been a by-product of the search for oil, domestic oil prices have undoubtedly affected gas reserve additions as well. Despite state conservation regulation limitations on output and the Mandatory Oil Import Program, domestic oil prices (measured in constant dollars) declined throughout the 1960s, as did prices for natural gas. (See Table 4.) The proportion of domestic oil (and gas) potential resources that it seemed profitable to explore and develop was therefore progressively reduced throughout the 1960s.

Table 4

CRUDE OIL AND NATURAL GAS FIELD PRICES
(1958 U.S. dollars)

Year	**Crude Oil** (price per barrel)	**Natural Gas** (price per Mcf)[a]
1963	2.69	15.5
1964	2.65	15.2
1965	2.58	15.1
1966	2.53	14.7
1967	2.48	14.4
1968	2.44	14.1
1969	2.41	13.7
1970	2.35	13.3
1971	2.38	13.5
1972	2.32	14.0[c]
1973	2.51[b]	13.2[b,c]

[a] Average price paid to natural gas producers by interstate pipelines.

[b] Controlled price.

[c] Average price paid by major interstate pipelines.

Source: Data on crude oil supplied by Hughes Tool Company; Natural gas statistics from a report by the Federal Power Commission, U.S. Senate, Committee on Interior and Insular Affairs, *Natural Gas Policy Issues and Options* (Washington, D. C.: Government Printing Office, 1973), p. 220.

Another factor responsible for the shortage of natural gas was the low rate of federal lease offerings in areas of prospective discovery. The most promising domestic acreage for hydrocarbon discovery and production is now the Outer Continental Shelf (OCS). Yet federal lease sales on the OCS have been insufficient to offset the depletion of promising onshore acreage.

The number of oil wells drilled declined from the record high of 31,567 in 1955 to 11,306 in 1972, thereby affecting the magnitude of reserve increases of associated gas. Since 1962, when the number of gas wells completed peaked at 5,848, gas well drilling generally declined through 1971 when 3,830 wells were completed (Table 5).

Table 5

U.S. WELL DRILLING STATISTICS BY CLASSIFICATION [a]

Year	Oil	Gas	Dry	Total
1955	31,567	3,613	20,742	55,922
1956	30,730	4,543	21,838	57,111
1957	28,012	4,620	20,983	53,615
1958	24,578	4,803	19,043	48,424
1959	25,800	5,029	19,265	50,094
1960	21,186	5,258	17,574	44,018
1961	21,101	5,664	17,106	43,871
1962	21,249	5,848	16,682	43,770
1963	20,288	4,751	16,347	41,386
1964	20,620	4,855	17,488	42,963
1965	18,761	4,724	16,016	39,501
1966	16,447	4,167	15,770	36,384
1967	15,329	3,659	13,246	32,234
1968	14,331	3,456	12,812	30,599
1969	14,368	4,083	13,736	32,187
1970	13,020	3,840	11,260	28,120
1971	11,858	3,830	10,163	25,851
1972	11,306	4,928	11,057	27,291
1973	9,902	6,385	10,305	27,602
1974 [b]	5,738	3,599	5,381	15,255

[a] Excludes service wells.

[b] First half 1974.

Source: Figures for the period 1950 to 1965 derived from the American Association of Petroleum Geologists and the *Oil and Gas Journal;* for 1966 to 1972, from the American Petroleum Institute and the American Association of Petroleum Geologists.

(A total of 4,928 gas wells was drilled in 1972, the increase reflecting a number of factors to be discussed subsequently.)

There has also been a sharp decline in all categories of drilling from the levels achieved in 1957. Total new field wildcat wells drilled in 1971 were 44 percent below the 1957 level. Over the same 1957 and 1971 period, all other categories of wells drilled in known fields declined by 51 percent. Natural gas well completions resulting from exploratory drilling increased through 1959, but since then have been in a long-term decline.[12] The success percentage for all exploratory

Table 6

THE PERCENTAGE OF SIGNIFICANT GAS FIELDS IN ALL GAS DISCOVERIES, MEASURED AFTER SIX YEARS

Year Drilled	Percentage
1945	52.69
1946	43.84
1947	44.90
1948	46.00
1949	50.89
1950	41.12
1951	47.10
1952	45.95
1953	32.70
1954	32.23
1955	24.62
1956	34.81
1957	42.31
1958	39.29
1959	20.13
1960	33.76
1961	18.59
1962	27.05
1963	23.26
1964	23.45
1965	23.21

Source: American Association of Petroleum Geologists *Bulletin*, July 1972.

[12] *Natural Gas Policy Issues and Options*, Senate Interior and Insular Affairs Committee, 1973, Table II-C, p. 218.

drilling has declined from a high of 20 percent in 1959 to a low of 15 to 16 percent.

From 1966 through 1972, the number of successful exploratory gas wells drilled declined in nearly every producing area. The average number of successful wells drilled in the years 1970 and 1972 was 42 percent less than in 1966 in the Southern Louisiana area and 54 percent less than in 1966 in the Permian Basin area. Of all the producing regions, only the Rocky Mountain area has shown an upward trend in successful exploratory gas well drilling.

Moreover, even if exploratory drilling and discovery of fields had continued at the peak levels of the 1950s, additions to gas

Table 7

NEW NONASSOCIATED NATURAL GAS DISCOVERIES AND NEW CONTRACT PRICES IN THE UNITED STATES, 1953 TO 1969

Year	New Nonassociated Discoveries (Tcf)	New Interstate Contract Price (cents per Mcf)	New Contract Price in 1958 Dollars
1953	8.5	13.3	15.1
1954	11.0	11.7	13.1
1955	6.8	14.4	15.8
1956	13.5	14.8	15.7
1957	13.8	16.9	17.3
1958	12.3	18.6	18.6
1959	10.3	18.4	18.1
1960	9.3	18.2	17.6
1961	7.0	17.9	17.1
1962	6.8	17.5	16.5
1963	9.4	17.0	15.9
1964	5.4	16.2	14.9
1965	7.1	17.4	15.7
1966	5.4	17.4	15.3
1967	3.4	18.6	15.8
1968	1.2	19.0	15.5
1969	1.7	19.7	15.4

Source: Patricia E. Starratt and Robert M. Spann, "Alternative Strategies for Dealing with the Natural Gas Shortage in the United States," in Edward W. Erickson and Leonard Waverman, eds. *The Energy Question: An International Failure of Policy* (Toronto and Buffalo: University of Toronto Press, 1974), vol. 2, "North America," p. 31.

reserves would have declined because the size of the average find has declined and the proportion of finds which are of significant size has declined (Table 6). With prices failing to increase as rapidly as costs, the declining size of finds has meant a declining prospective return on exploratory investment (Table 7). This has reduced the incentive for exploratory drilling and discovery.

3

EFFECTS OF FIELD PRICE REGULATION

Producers' decisions depend on regulated prices in several ways. If a producer is contemplating a large investment in exploration and development, the current price ceiling and the regulatory (and political) climate suggest to him what he may expect in future years when the gas comes onstream. And, where gas is to be produced from federal lands (from which it must be sold interstate), FPC ceilings have a direct bearing on future profitability. The "go" or "no-go" decision on various drilling prospects will be influenced therefore not only by current price ceilings but also by projections about the future regulatory and political climate. In the absence of field price regulation, the producer can more safely assume that more marginal prospects will be worth drilling.

Investment in natural gas exploration and development has accelerated in the last three years in response to rising prices in intrastate markets and the anticipation of deregulation (or, at least, of higher prices for gas sold interstate). This new upward trend has continued through the first quarter of 1974; gas well completions totaled 1,800 as opposed to 1,392 completions in the first quarter of 1973.[1]

Even if the gas can be sold intrastate, FPC price ceilings will indirectly affect price levels in the unregulated sector over the long term. If interstate prices are kept low in the future, or regulation is extended to the intrastate market, profitability expectations will drop and investment will decline. Expectations play a key role in investment decisions, and regulation in and of itself is a deterrent to investment.

---

[1] John N. Nassikas, chairman, Federal Power Commission, statement before the Senate Commerce Committee, 20 August 1974, p. 5.

Uncertainty

Large investments are deterred by uncertainty. This has been and remains the keynote of the regulatory scene. Years before new supplies can be delivered to the customer, the producer must make the decision to invest his capital to drill. A long period of uncertainty characterizes the regulatory process because of the difficulties in determining price ceilings and the inevitable parade of cases through the courts. The amount of time involved before final determination in area rate cases is reached is illustrated by the following list of cases in the major areas of production:

Area Rate Case	*Date started and status*
Permian I, II	1961–65, 1970, Commission decision,, Aug. 7, 1973.
Southern Louisiana I, II	1961–74, Commission affirmed by the 5th Circuit Court, April 1973; Supreme Court affirmed, June 10, 1974.
Hugoton-Anadarko	1963–71.
Texas Gulf Coast	1964– , remanded to Commission by U.S. Court of Appeals for the District of Columbia Circuit; Supreme Court remanded to District of Columbia Circuit Court June 17, 1974.
Other Southwest	1967–71, still on judicial review, Commission affirmed by 5th Circuit Court but granted power to reopen with respect to old gas; Supreme Court denied certiorari June 17, 1974.

The FPC is making efforts to reduce regulatory lag. In July 1971, the commission announced its decision to use rule-making techniques in setting price ceilings. The rule-making procedure has three steps. First, the commission issues its Notice of Proposed Rulemaking, inviting comments from interested parties. Second, the comments are submitted (they may include cost studies, market value studies, and so forth). Opportunity for answering comments is provided in some cases. Third, the FPC issues its ruling, either prescribing a specific price ceiling or a procedure whereby the standards for price rulings are established.

Specific examples of FPC rule making in the rate-making process include (1) the setting of rates for the Appalachian, Illinois Basin, and the Rocky Mountain areas and (2) the procedures for special relief and emergency sales at above ceiling prices, as well as for the Optional Pricing Procedure and Small Producer Exemption. Commission action in the Rocky Mountain case actually began in July 1971, but the commission did not issue its opinion until April 1973. On the average, it appears that, even using the Optional Pricing Procedure (wherein buyer and seller come to the FPC with a proposed price level), the FPC has been unable to issue decisions until after eight or nine months.

Beyond the regulatory lag, there is the inevitable judicial delay. Litigation in the past further extended the time period before decisions become final (often for more than a decade). Even today, after the commission issues a decision, final determinations may still take three years. No matter what standard is set for the determination of appropriate price levels or ceilings, moreover, litigation can be expected even if the burden of proof is shifted to the intervenor. Regulatory and judicial delay together currently deter investment in domestic exploration and development.

Producers have also found that previously approved rates may be reduced. Under its statutory rate investigation authority, the FPC is empowered to reduce a rate it once approved (or a rate approved by members of an earlier commission), and this power has been used. In late 1960, the Southern Louisiana area guideline rate was set at 23.25 cents; many producers undertook exploration programs on the basis of this price. Yet the level at which certificates were issued was then reduced to 21.25 cents, then to 20.625 cents, to 20 cents, and, finally, eight years later, the producers were told that the acceptable price would be no more than 18.5 cents per Mcf. Even though the prices allowed have been rising in the last three years, there is always the possibility that a future commission might roll back currently acceptable rates.

Uncertainty on prices to be allowed has had a pernicious effect on supply. For over a decade, producers have not known what prices would ultimately be allowed for the gas they sold. Where prices have been reduced below contract levels, producers have been obliged to pay out enormous sums in refunds. Producers today probably believe prices will continue to rise, but judicial, administrative or legislative changes could well falsify that belief.

Imprecision

At best, FPC costing techniques produce only approximate results, with a tendency to err on the low side in the fixing of ceiling rates. Producers make decisions according to the projected overall costs of a specific project and the return the project is likely to yield a few years hence. They cannot keep separate books on oil costs, or gaseous liquids costs, inasmuch as any attempt to do so would require the use of arbitrary cost allocation formulas that make for economic nonsense.

The utility rate-making formulas used by the FPC in the setting of field price ceilings are inappropriate to the economic conditions of natural gas exploration, development and production. For these formulas to be used, cost of service must be set. It can only be set by arbitrary methods because some 40 percent of the expenditures of natural gas production are jointly incurred on the production of other hydrocarbons. Even in gas reservoirs not associated with oil, the presence of liquid hydrocarbons (butane, propane, ethane and pentane) requires arbitrary cost allocations. No matter what technique is used to separate joint costs, the result is arbitrary because the costs are inseparable.

The effort to determine producer price ceilings based on cost has come under particularly sharp attack from some economists:

> The way those presenting data and calculations of average cost of an Mcf of gas in area hearings switch back and forth among sources, the way they make joint cost allocations, and the way they apply "factors" represent pseudoscience to a degree it would be difficult to equal. At the final determination, the examiner's fumbling among the numbers and making a cafeteria-style selection from those presented, plus the Commission adding a few delicate adjustments of its own, make the whole thing nothing short of ludicrous. . . .[2]

Several observers have noted that at least some FPC price determinations have as their basis the price level that "feels right" to the person charged with determining the "right" price:

> In the allocation situations—especially in Permian—every argument is justified in heavy legalisms, any of which is "reasonable" to a particular view, and even as truly an objective approach taken to choose any particular one as possible

[2] Clark A. Hawkins, "Structure of the Natural Gas Producing Industry," *Regulation of the Natural Gas Producing Industry*, ed. Keith Brown (Washington, D. C.: Resources for the Future, Inc., 1972), p. 165.

would find it at best, in the majority of cases, only marginally better than those rejected. It is not absurd to feel that the examiner in each of these major cases arrived at an imprecise but definite idea of what a "fair" price was during the course of the hearing, and then saw to it that very close to this ("fair price") was generated from the abundance of numbers at his disposal when the decision was to be rendered. The major portion of the time spent, then, in preparing the decision would be spent in justifying, on legal and logical grounds, the numbers chosen. It is easy to imagine how difficult it would be to keep from doing so.[3]

The Supreme Court in Permian I specifically acknowledged the problems inherent in allocating joint costs of gas:

> Economists have described these difficulties with repetitive pungency. "To make laborious computations purporting to divide [such] costs is 'nonsense on stilts,' and has no more meaning than the famous example of predicting the banana crop by its correlation with expenditures on the Royal Navy."[4]

An example of the differences in "cost" estimations may be found in the recent cost determination for gas from Southern Louisiana. In 1971, the FPC staff estimates were 24 to 26 cents per Mcf. In 1973, the staff concluded that the range of cost was between 28.27 cents and 36.58 cents per Mcf. However, too much reliance may have been placed on pre-1969 data in computing those costs. The staff employed productivity averages for 1947 to 1971 (580 Mcf per foot drilled) for the low side of the cost range and averages for 1966 to 1971 (525 Mcf per foot drilled) for the high range. Average productivity for 1966 to 1972 was 500 Mcf per foot drilled, and for 1969 to 1972, 350 Mcf per foot drilled. With the updated figures (and taking into account the time value of money as well as other factors), the United Distribution Companies had computed average costs for nonassociated gas in Southern Louisiana to be in the range of 42.17 to 67.96 cents per Mcf (with a 16, rather than 15, percent rate of return). Ironically the 26 cents per Mcf area rate for Southern Louisiana was still being litigated as the "correct" price ceiling.

Then too, there is a problem involved in using average or aggregate cost numbers: these figures represent *no* particular company's cost. Price ceilings based on FPC estimates of average cost in fact make it likely that roughly half the gas currently being produced

---

[3] Ibid., p. 166.

[4] 390 U.S. Stat. at L. 804, n. 80.

costs more to develop than the producers can receive for it. In other words, if the ceiling price is 35 cents, none of the gas that cost more than 35 cents to produce would have been developed—had the producer known in advance what the gas would cost and that the ceiling price would be 35 cents. Moreover, price ceilings for what is determined (by date of contract or well completion date) to be "new" gas are based on the same data that were used as the basis for previous price ceilings. With utility rate-making formulas, prospective costs cannot be considered, and "new" gas price ceilings are always based on past data.

In fact, cost-price circularity is built into the regulatory system. Price ceilings limit exploration to those reserves the producer believes can be economically developed in the future. To the extent that price ceilings remain low, and to the extent these price ceilings influence drilling, only low-cost reserves will be developed. Accordingly, price ceilings based on "cost" will remain low—since only low-cost wells were drilled because of anticipated low-price ceilings—even though geological uncertainty guarantees that some gas will fall in the high-cost category. Since the cheapest gas is always developed first where possible, exploratory effort will become more and more restricted by cost-based ceilings over the long term.

Differential Pricing for "Old" and "New" Gas

The "two-tier" approach to pricing gas is based on an arbitrary distinction between old and new gas. Most of the gas that has been found in the last decade has in fact been the result of extensions of old fields, not of discoveries in new fields. The regulatory practice of paying lower prices for old gas than for new has led to resource misallocation. Gas is being left in the ground which otherwise could have been produced. Where the ceiling price fails to cover the costs that are incurred in keeping an old well productive, the well will be abandoned as "uneconomic." These costs of keeping an old well productive include the use of compressors, reconditioning the well, and alterations to drill deeper or shallower reservoirs in the same field. In the words of M. A. Adelman,

> the payment of lower prices for so-called old gas discourages the more intensive development of old pools and the search for new pools in old fields. It makes no economic sense to leave these unexploited if we are willing to pay a higher price for gas, as evidenced by the higher prices for the so-called new gas. It is also senseless and unproductive to

> have some purchasers get a windfall in the shape of the cheaper old gas as against those who must pay higher prices for the new gas. Windfalls to the owners of unusually good reservoirs do serve an economic purpose—to encourage fresh investment in new pools.[5]

Resource Misallocation between Interstate and Intrastate Markets

While it is not widely recognized by legislators from consuming states, the chief beneficiaries of FPC regulation have been intrastate industrial consumers. Federal regulation of the price of gas sold interstate depresses the price of natural gas in all markets. A low field price encourages industrial use in areas close to the source of supply. In the words of Professor Edmund Kitch,

> The second reason that the present regulation has encouraged the industrial uses of natural gas is related to the regional distribution of those industrial uses. The west-south-central area is composed of the States of Texas, Louisiana, Arkansas, and Oklahoma. In these States natural gas is largely supplied by the intrastate market. This area is the most intensive natural gas consuming area in the Nation. The area consumes 34 percent of the natural gas produced in the Nation. Ninety-one percent of the gas consumed in this region is consumed industrially. Put another way, 40 percent of all natural gas in the United States which is consumed industrially is consumed in the west-south-central area. By holding down the price of natural gas within the region, the Federal regulation has effectively acted as a subsidy to this industrial market, and therefore as a subsidy to the industrial growth of the Southwest. The only practical way to reduce the industrial use of gas within the Southwest is to raise the price of gas in that region. . . . Put another way, the *residential gas consumer of the Pacific Coast, upper Midwest and the East Coast is prevented by Federal law from paying 10 to 15 percent more for his gas, thereby making gas in the American Southwest 50 percent cheaper than it would otherwise be and subsidizing the movement of industry from the consumer's home region to the Southwest.* [Emphasis added.] [6]

---

[5] M. A. Adelman, statement before the Senate Interior and Insular Affairs Committee, 25 February 1972 (Serial No. 92-22), p. 56. On 12 April 1973, the FPC issued two orders to grant special relief where factors such as reduced pressure made continued production uneconomic at existing rates.

[6] Edmund W. Kitch, "The Shortage of Natural Gas," *Occasional Papers from the Law School*, The University of Chicago, 1 February 1972, pp. 10-11.

Further evidence of the misallocation of resources from interstate to intrastate consumers was provided by past Chairman Joseph C. Swidler of the New York State Public Service Commission. In the five-year period from 1966 through 1970, interstate pipelines lost heavily to intrastate buyers in bidding for new gas supplies from the Permian Basin area. In fact the index of relative position moved from 5 to 1 (in favor of interstate buyers) to 1 to 10, a deterioration of 50 times:

> Not only did relative position change drastically, but also the absolute amounts. In the 5-year period (assuming the figures for the second half of 1970 were consistent with the first half of the year), there was a decline from 149 billion cubic feet to 20.6 or a fifth as much as in 1970. *While a crisis faced the consumers in the interstate market, the sales for 1970 to intrastate buyers were apparently at least seven times the 1966 sales*—almost four times total 1966 sales in the first six months of 1970 alone. . . . [Emphasis added.] [7]

Why were the intrastate consumers getting almost all the new gas by 1970? One explanation is that leasing of offshore federal lands took place at a slow and erratic pace (gas produced on federal lands must by law be sold in interstate commerce). But the reason for the interstate pipelines' inability to obtain sufficient supplies from non-federal areas can only be attributed to price ceilings set by regulation. To win a contract, when supply was beginning to lag behind demand, all the intrastate buyer had to do was bid a higher price than the interstate pipeline was permitted to bid by regulation:

> Prices offered by intrastate buyers for the new gas in this area rose from 17 cents per Mcf in 1966 to 20.3 cents per Mcf in 1970, and toward the end of 1970, the intrastate pipelines bought more than 200 billion cubic feet of reserves at initial delivery prices of 26.5 cents per Mcf. At the same time, prices paid by interstate pipelines could not exceed the regulated ceiling and therefore remained between 16 and 17 cents per Mcf. *The inescapable conclusion is that the interstate pipelines were simply outbid.* [Emphasis added.] [8]

Statistics on sales to ultimate consumers also support the view that

---

[7] Joseph C. Swidler, statement before the Senate Interior and Insular Affairs Committee, "Natural Gas Policy Issues: Part I," pp. 298-299.

[8] Stephen Breyer and Paul W. MacAvoy, "The Natural Gas Shortage and the Regulation of Natural Gas Producers," *Harvard Law Review*, vol. 86, no. 6 (April 1973), p. 979.

the result of FPC policy in the 1960's was to deplete the gas reserves of interstate home consumers in favor of the demands of intrastate industrial consumers to whom sales were unregulated. . . . The percentage of gas sold by pipelines and distributors to residential users declined by 1.6 percentage points between 1962 and 1968. This decline was caused in large measure by a substantial increase in industrial sales by unregulated intrastate pipelines and by producers themselves. Between 1962 and 1968, total industrial consumption of natural gas increased 43.5 percent, while intrastate pipelines and distributors increased their industrial sales by almost 62 percent. Moreover, of the increase in industrial consumption, more than half can be attributed to sales by intrastate pipelines and distributors, while less than 13 percent is accounted for by direct industrial sales of the interstate pipelines. The remaining 37 percent of the increase was the result of direct sales by the producers. [Emphasis added.] [9]

Low Prices and Low Supply

The imbalance between the quantity of natural gas supplied at current prices and the quantity demanded can be seen as a virtually inevitable result of the low regulated producer prices and the delays caused by regulation and litigation. This is the conclusion of most economists who have never been associated with the industry or its regulation but who have studied the relevant data over a period of years. While the slow and erratic pace of federal offshore lease sales has contributed to the imbalance, reserves from nonfederal lands onshore have provided much of the gas developed in the past and could account for more than half the gas developed in the future.[10] If prices had been allowed to rise over time in response to changing economic circumstances, more onshore gas would now be available.

The purpose of regulation, from a regulator's point of view, must be to prevent interstate pipelines from paying market prices for natural gas. The end result of such regulation would have to be prices that are lower than they would have been under free market conditions. Since free market prices are depressed by surpluses and raised by shortages, thereby being moved to levels which balance the

---

[9] Ibid., p. 977.

[10] U.S. Geological Survey, Department of the Interior, *Revised Petroleum and Natural Gas Resource Estimates*, News Release, 26 March 1974, p. 5.

quantities demanded and supplied, any price held below the free market level prevents the elimination of shortages.

To the extent that prices have been lowered by regulation, consumers who are able to get gas have benefited from lower cost gas. One study concluded that wellhead prices were on the average $.06 per Mcf below market-clearing levels in the 1960s. One might therefore multiply the average annual production of regulated gas from 1962–68 (11 Tcf) by this $.06 and conclude that producer regulation saved consumers about $660 million annually. Yet, one should exercise caution in using this popular device:

> *Such a calculation contains heroic assumptions and oversimplifications.* For one thing, it assumes that every cent of price reduction at the wellhead was passed through to ultimate consumers; in light of the fact that sales by retail distributors are intrastate and therefore subject only to State regulation, the assumption may not be valid. For another thing, had producers received a higher price, at least some of their additional revenues would have been taxed away and, therefore, indirectly returned to consumers anyway. Nonetheless, even assuming that the entire 6 cents per thousand cubic feet was returned to consumers who actually received gas, *we still doubt that this benefit outweighed the losses arising from regulation, even from the point of view of the consumer class itself.* [Emphasis added.] [11]

For various reasons, moreover, it can be said that field price regulation may have little to do with price increases to the ultimate consumer, especially the residential consumer. It can reasonably be argued that field price regulation has not only depressed supply but has also failed in protecting the consumer:

> The average price of gas at the wellhead has gone up about 5.5¢/Mcf and at the point of consumption about 15¢/Mcf since 1954. Thus, even though producers were regulated, the price to the consumer went up nearly three times as much as the wellhead price. The F.P.C. has undoubtedly exerted some restraining influence on field prices, even though the extent of the increase in wellhead prices, had there been no regulation, is problematical. *It could be argued that the F.P.C.'s regulatory policy on field prices is such a small part of the consumer's price, but at the same time may have adversely affected the rate at which natural gas supply is*

[11] Breyer and MacAvoy, "The Natural Gas Shortage and the Regulation of Natural Gas Producers," p. 980.

forthcoming to meet the demand for it in the energy requirements of the remainder of this century. [Emphasis added.] [12]

Statistics suggest that the prices various consumers pay are largely determined at the local level. Consolidated Edison charges the New York residential consumer $7.76 per Mcf for the first block (first 300 cubic feet or less) of natural gas, for example, whereas the industrial consumer on an interruptible contract is charged only $0.70 per Mcf for the first block (the first 15,000 Mcf or less).[13] That field price regulation had a depressing effect on supply appears obvious from Table 6. Real prices for new natural gas supplies fell throughout the 1960s, as did the rate of new natural gas discoveries:

> The real wellhead price of natural gas fell 15 percent in the first nine years of field price regulation. Discoveries of natural gas peaked at approximately the same time as did real new contract prices. . . . The current shortage is a direct result of the reduction in the rate of new natural gas discoveries during the 1960s, and this low level of natural gas exploration and discovery was due to FPC ceilings.[14]

[12] Hawkins, *The Field Price Regulation of Natural Gas,* p. 212.

[13] Letter of 16 January 1974 by Joseph Swidler, chairman, Public Service Commission of New York, containing supplementary answers to questions by Senator Dewey Bartlett, subsequent to testimony before the U.S. Senate Special Subcommittee on Integrated Oil Companies, Committee on Interior and Insular Affairs, p. 2.

[14] Patricia E. Starratt and Robert M. Spann, "Alternative Strategies for Dealing with the Natural Gas Shortage in the United States," in Edward W. Erickson and Leonard Waverman, eds. *The Energy Question: An International Failure of Policy* (Toronto and Buffalo: University of Toronto Press, 1974), vol. 2, "North America," p. 141.

4
COMPETITION IN FIELD MARKETS

The natural gas producing industry is obviously neither purely monopolistic nor purely competitive. Estimates of the total number of domestic oil and gas producers range up to 30,000. Natural gas pipelines nationwide number 137. While there are clearly many independent sellers and independent buyers in field markets, in particular areas for limited periods of time, monopoly (sellers') power may exist on the part of the producers, and monopsony (buyers') power may exist on the part of the pipelines.

Academic Studies

Many studies have been conducted over the years in an effort to determine whether or not buyers or sellers typically possess sufficient market power to make competition in field markets unworkable. It is not an exaggeration to say that almost all recent academic studies have found that natural gas field markets would be workably competitive in the absence of field price regulation.

> The conclusion reached in most studies is that while the industry does not fit the carefully drawn competitive model on either the demand or supply side, it is workably competitive. . . .[1]

Moreover, the relevant data suggest that pipelines generally have more market power than the producers. Some of the experts have

[1] Milton Russell, "Producer Regulation for the 1970's," *Regulation of the Natural Gas Producing Industry*, ed. Keith Brown (Washington, D. C.: Resources for the Future, Inc., 1972), p. 220.

concluded on the basis of their studies that producer market power was virtually absent.

> Some of the markets are seen to exhibit price and sales behavior markedly similar to the theoretical monopsony pattern, and none indicates behavior characteristic of the theoretical monopoly pattern.[2]

In the words of Keith C. Brown,

> While it is clear that neither on the demand side nor supply side does the market fit the economist's ideal of price competition, most economists feel the evidence warrants a verdict of effective or workable competition on the supply side, that is, absence of market power on the part of producers with perhaps some monopsony (purchase market power) in some special areas on the demand side.[3]

Producers do have pipelines to contend with in setting prices. Pipeline monopsony has been in evidence in the past and is still in evidence today:

> The available evidence shows, for example, that the four largest production companies provided only 37–44 percent of new reserve sales in the West Texas-New Mexico producing area, 26 to 28 percent in the Texas gulf region, and less than 32 percent in the midcontinent region—all in the 1950–54 period just before the *Phillips* decision [wherein the Supreme Court issued a ruling containing the ambiguous language that was interpreted by some to indicate that the Federal Power Commission should regulate field prices for natural gas sold interstate.] These levels of concentration on the supply side of the market for new reserves were all less than half the concentration on the demand side, accounted for by the four largest pipeline buyers in each of these regions. *Power to control new contract prices probably did not exist on either side of the market, but if the scales tipped at all, then surely the balance lay with the pipeline companies rather than with the producers.*
>
> Of course one can still argue that despite its apparently competitive structure the producing segment of the industry has behaved noncompetitively. Certain proponents of producer regulation have pointed to the rapid rise in the field price of natural gas between 1950 and 1958 as evidence of

---

[2] MacAvoy, *Price Formation in Natural Gas Fields*, p. 9.

[3] Keith C. Brown, "Introduction," *Regulation of the Natural Gas Producing Industry*, p. 2.

such noncompetitive performance. *But economic studies of the markets for new contracts suggest that anticompetitive producer behavior did not cause this price increase.* During the early 1950's the presence of only one pipeline in many gas fields effectively allowed the setting of monopoly buyers' (monopsony) prices for new gas contracts, thus often depressing the field price below the competitive level. During the next few years, several pipelines sought new reserves in old field regions where previously there had been such a single buyer. This new entry of buyers raised the field prices to a competitive level from the previously depressed monopsonistic level. In short, competition—not market power—accounted for much of the price spiral that has been claimed to show the need for regulation.[4]

In testimony before the Senate Interior and Insular Affairs Committee's Subcommittee on Integrated Oil Operations, 13 December 1973, Professor Paul W. MacAvoy stated that

our review . . . shows remarkably strong diversity of sources of new reserves, with low levels of concentration in the provision of these reserves in Gulf Coast markets, the Mid-Continent markets, and the Permian Basin region of West Texas. Diversity is reduced when account is taken of the practice of producers to take part in joint ventures. In the extreme case, they would not compete with each other in the sale of new reserves; assuming the extreme case, companies systematically bidding together can be termed one company so that the shares of the largest firms are increased in some cases by appreciable amounts and in other cases scarcely at all.

Overall, the number of producers with significant alternative sources of supply greatly exceeds the number of buyers that are valid sources of demand in the interstate markets. Concentration on the buying side of these markets is in most regions twice as high as on the selling side. When consideration is given to "joint ventures" among the buyers —that is, to the common practice of exchanging gas among themselves—the largest four buyers in any one of the three important markets [Gulf Coast, Mid-Continent and Permian Basin] control more than 80% of total purchases in those markets in the late 1960's.[5]

[4] Breyer and MacAvoy, "The Natural Gas Shortage and the Regulation of Natural Gas Producers," pp. 944-949.

[5] Paul W. MacAvoy and Robert S. Pindyck, statement before the Senate Interior and Insular Affairs Committee, 31 December 1973, pp. 2-3.

In other words, the monopsony power of the interstate pipelines—the power to depress prices—as measured by concentration ratios, is still much greater than the monopoly power of the producers.

Key Flaws in the "Market Power Theory"

One-sided Analysis. Two prominent spokesmen espouse the view that competition in field markets is unworkable (or, indeed "nonexistent") and that more regulation is needed. Close scrutiny reveals significant flaws, however, in the work that supports their conclusions. The testimony of both David S. Schwartz and John W. Wilson is analytically deficient in the areas of concentration, regulatory restraints, long-term profitability, joint bidding, ease of entry, and the ability to control prices.

Inaccurate concentration ratios. The factors which determine their conclusions include concentration ratio data, but here as in other areas their analysis is inaccurate. The testimony of the FPC Staff Witness John Wilson on the concentration of "new" gas supplies was rejected by the Commission in the Belco case (Belco Petroleum Corp., Docket No. CI73-292, FPC). Wilson had introduced evidence indicating that in offshore Louisiana, two producers accounted for 56 percent of the estimated first year volumes under certified 1972 contracts; four producers accounted for 68 percent, and eight producers accounted for 85 percent.

But offshore Louisiana should not be adduced to support the "Market Power Theory" of the natural gas shortage without additional consideration. First, the natural gas market is national. While about half the remaining gas to be found will be found offshore, about half will be found onshore. Even offshore gas is found in other areas in the Gulf of Mexico besides offshore Louisiana. Second, through pipeline interchanges and the development of extensive gathering lines, pipelines are not restricted as they were in the past in obtaining supplies. Third, the early sales figures by a few producers in a limited area for a brief time do not accurately reflect a real-world situation. Further development and sales will be made in the Gulf of Mexico, by the seventy-four producers who obtained leases in the federal lease sales of 1970 and 1972. The ranking of the sellers in ownership of uncommitted reserves differs substantially from their ranking in total sales:

> For example, the largest seller (in terms of total sales) was fourth in ownership of uncommitted reserves as of June 30,

1973; the second largest seller was eighth; and the tenth largest seller ranked 28 in ownership of uncommitted reserves for total Continental United States. Only two of the eight largest sellers ranked among the eight largest holders of uncommitted reserves in the Offshore Federal South Louisiana area at June 30, 1973 and four of the eight largest sellers had no uncommitted reserves in this area. The largest seller ranked number 15 in holdings of uncommitted reserves in this area and the next three sellers, ranked by total sales, had no uncommitted reserves in the area. . . .

None of the eight largest sellers was among the eight largest holders of uncommitted reserves in both the Offshore Federal South Louisiana area and the Permian Basin which, together, account for more than half of the total uncommitted reserves reported. [In fact] a number of producers not among the first 20 in total sales rank relatively high in ownership of uncommitted reserves in some areas.[6]

Failure To Account for Regulatory Restraints. In the last analysis, debates about producer concentration ratios are largely meaningless at this time. (Paul MacAvoy and John Wilson agreed to this point in a debate before the Senate Commerce Committee.) Federal price controls distort the market structure, providing only tentative indications of what market structure and performance would be if there were no price controls:

The impact of price controls on the market is overwhelming —entry is suppressed, exit of companies is accelerated, and the market shares of continuing companies are more or less randomly determined by fortuitous finds of new gas reserves. The market structure that would exist without regulation of prices cannot be extrapolated from present market conditions—*any more than one can predict the behavior of markets for alcoholic beverages from structural supply conditions during the period of prohibition.*[7]

Failure To Consider Long-Term Profitability Indicators. Professor Edward J. Mitchell of the University of Michigan conducted a study of long-term profitability, as measured by rates of return to stockholders. He found that, from the 1953 to 1972 period, American

---

[6] Letter to Senator Henry M. Jackson from W. O. Senter, 6 December 1973, enclosing Foster Associates summary by Mrs. Celia S. Gody of data on uncommitted reserves submitted to the FPC in Docket No. R-405.

[7] MacAvoy and Pindyck, statement before Senate Interior and Insular Affairs Committee, p. 2.

petroleum companies were less profitable than the Standard and Poor's 500 corporations. Furthermore, oil- and gas-producing companies were significantly less profitable than refiners and international companies, particularly after gas price regulation began in earnest. During the period from 1960 to 1972, the 500 Standard and Poor's corporations earned 12.8 percent per year. Ten oil- and gas-producing companies averaged only 6.3 percent per year, far below the Standard and Poor's 500 and the 15 percent nominally allowed by the FPC.[8]

Failure To Test Joint Venture Hypotheses. As for the "interlocking relationship" assumed to exist because of joint bidding and joint ventures, it is common knowledge among economists that the effect of joint ventures in limiting competition diminishes sharply with the number of separate ventures and the different partnership arrangements. It may be noted that hypotheses on the probable rationale behind joint ventures (for risk-sharing or collusive purposes) can and have been tested.

Studies of the 1972 and 1973 federal offshore lease sales conducted by Professors Edward W. Erickson and Robert M. Spann revealed that the observed patterns of bidding partnerships were most consistent with the hypothesis that joint bidding in federal lease sales takes place for the purpose of risk sharing, not for collusion. Erickson and Spann found that the membership of bidding groups varies from year to year; that market shares for members fluctuate from one year to the next; that the most frequent bidding group is a combination of majors and smaller firms, but the majors also bid alone, as do smaller firms. Single firm bids are frequently made by the largest firms, but other firms are also successful single firm bidders, and bidding groups which contain a large number of firms are predominantly composed of smaller firms. They concluded that the incidence of joint bidding increased as the size of the firm decreased. There is a high incidence of joint bidding partnerships between unlike firms. Smaller firms used joint bidding ventures as a vehicle for entry into offshore activity.[9]

Joint bidding thus facilitates ease of entry; smaller independents can and do form independent and successful groups for bidding

[8] Edward J. Mitchell, statement before the U.S. Senate Subcommittee on Integrated Oil Companies, Committee on Interior and Insular Affairs, 21 February 1974, reprinted as Appendix B of Edward J. Mitchell, *U.S. Energy Policy: A Primer* (Washington, D. C.: American Enterprise Institute, 1974), pp. 89-103.

[9] Edward W. Erickson and Robert M. Spann, statement before the Senate Commerce Committee, 8 November 1973, pp. 6-7.

purposes in federal lease sales. Because of the number of large- and small-producer combinations and the frequency with which new groups are formed with different members, producers are able to become involved in many projects. As a result, competition is increased, not diminished. It is important also for it to be remembered that joint operating agreements expressly provide for separate marketing of production from joint ventures. In the last analysis, if the formation of these groups created monopoly power, one would expect to see this monopoly power reflected in a nonrandom bid price pattern in federal lease sales. No such pattern has been found.

Failure To Account for Ease of Entry with Higher Prices. An important measure of monopoly power is the condition of entry for new sellers. Advocates of the "Market Power Theory" argue that the bidding system in federal lease sales offshore (which does not account for onshore activity) prohibits independents from entering the market to compete with the largest producers. There are two flaws in this reasoning. First, such arguments ignore the depressing effect of field price regulation on entry. Independent producers were dropping out of the business long before federal lease sales were even scheduled, and before bids became as high as they are today. The most plausible explanation for the fall-off in drilling and production by independents is that domestic prices in the field fell further and further from covering costs. Regulation depressed prices in the interstate market and indirectly in the intrastate market. Until the advent of severe production shortages, the domestic production business was not profitable. Only producers protecting the value of sunk investments or in a position to use funds that would otherwise be taxed continued ambitious exploration and development activities. Low prices were the major disincentive to new entry into the market.

Second, ease of entry does in fact prevail. Studies by University of Arizona Professor Clark A. Hawkins led him to the following conclusions:

> Except for possibly the risk factor, the barriers to entry for natural gas production are substantially less than in most manufacturing. Even the risk case is not clear, since it may be an inducement to certain types of venture capital that would be in a high marginal tax bracket. The rise of mutual-type funds selling pieces in drilling ventures in recent years is one indication that the high risks themselves do not deter entry.[10]

---

[10] Hawkins, *The Field Price Regulation of Natural Gas*, p. 212.

Price increases since 1971 have encouraged entry, and entry has occurred. In the 1972 and 1973 federal lease sales, there were thirty-six successful single bids by independent producers smaller than the eight major producing companies. At least five of the independent producers were not in any way "interlocked" with the top twenty producers. Recently fifteen companies entering for the first time have had winning bids in federal offshore lease sales. The substantial entry of pipelines and affiliates in the production business further suggests ease of entry and competition among producers. Pipelines and affiliates obtained an interest in 53.5 percent of the leases awarded in the September 1972 federal lease sale and 54.3 percent of the leases awarded in the December 1972 lease sale.

Monopoly Power and the Ability To Control Prices. Except under circumstances of political motivation (as in the Mideast oil cutoff by OPEC in October 1973), the classical behavior of individual members of a cartel seeking to increase profitability is to increase capacity in order to enlarge the claims for their fair share of allowed production. It is inconsistent, therefore, to claim on the one hand that the major domestic producers operate "like a cartel" and to accuse them, on the other hand, of "refusing to develop reserves." Such a statement, while politically appealing, makes no economic sense.

Finally, even assuming that producers do have sufficient market power today to raise prices above competitive levels, it is important to understand that the power to keep prices at inflated levels would be short-lived. In a free market, as prices rise, the rate of demand decreases and the total quantity supplied increases. Over time, then, prices will fall from initial levels reached immediately subsequent to decontrol as the supply/demand circumstances change. Thus, even if leading producers sought to alter their behavior in a way leading to noncompetitive results, "the ambit through which such behavior could function, if it existed, is restricted both in magnitude and duration." Even if no new independent producers entered the field, moreover, the top eight or twenty producers have never been described as having the power to control the output of the 4,350 producers (below the top-twenty category) currently selling gas in the interstate market.[11]

[11] Russell, "Producer Regulation for the 1970's," p. 220.

5
POLICY ALTERNATIVES

Regulation: "Cost" Standard

This alternative is not viable. The FPC has been regulating prices on this basis since 1960. Although touted by "consumer advocates," history and economic analyses offer persuasive arguments that such regulation cannot be effectively administered and does not serve the consumer's interest in supply or, ultimately, in price. While such regulation has benefited lawyers and consultants, it is the primary cause of the natural gas reserve and production shortage and is also the chief source of the drive for expensive gas imports from Algeria and Russia.

This alternative is fraudulent in that gas production costs cannot even be estimated with reasonable accuracy. In response to Dr. John W. Wilson's contention that this view is "unfounded," Frank C. Allen, chief of the FPC's Bureau of Natural Gas, made the following comments:

> In a number of area rate proceedings, the FPC has determined the average costs of producing gas, but these determinations have only been made after considering testimony from different witnesses representing a wide range of opinion as to the proper level of the different cost elements. It should be noted that only a portion of the cost of finding, developing and producing new gas well gas is relatable to specific gas volumes. . . . The remainder is composed of exploratory expenses that is never relatable to specific gas volumes and return on investment. Where oil [or liquids] and gas are produced jointly from the same lease, the cost allocations between the two products are very difficult and complex.[1]

[1] Frank C. Allen, comments on "Policy Issues and Options Affecting Natural

Whereas Dr. Haskell Wald, chief of the FPC's Office of Economics, stated that there is no evidence that the Federal Power Commission sets rates on the low side, Allen commented:

> This statement is incorrect inasmuch as in establishing the first area rate price ceiling the Commission used the average price of gas for an historical test period. Thus, the cost of gas was unavoidably low for many producers for two reasons. One, when the average cost of gas is used to set the price ceiling, by definition, some producers will have costs above the average, others below the average. Those with costs above the average would thereby have costs in excess of the area price ceiling they were allowed to charge. Two, in Permian I an historical test year was used thereby estimating costs and rates for the future on the basis of past experience. Accordingly, *the area rate reflected only past costs and if a producer's cost increased he would not recoup those costs even if a purchaser would pay the higher price. Thus the procedure developed for setting area rates had the undeniable effect of pricing certain producers out of business. . . .* [Emphasis added.] [2]

Cost-Based Regulation Is Arbitrary. Cost-based regulation operates as arbitrarily as relating the gas price to a price number for another fuel on the basis of BTU equivalency, and deciding that the resulting price level should be the ceiling for "new" or "old" gas. Using past cost data to develop price ceilings instead of economic factors merely guarantees that prices will be too low to elicit adequate new supplies.

Cost-based regulation severely limits supply. Only the lowest cost projects may be profitably developed, and low price ceilings result. Since costs are determined on the basis of past data, ceilings based on cost alone cannot reflect prospective costs. Nor is the incentive to drill increased with a few cents per Mcf added for noncost factors. The cost figure is simply too low to start with and should therefore not be included as a regulatory standard in any circumstance. Even the FPC staff has argued (in Docket R-389A) that

> to induce producers to find more gas in the period ahead, the Commission must provide an economic incentive for them to explore a wider range of prospects and to be willing to incur higher exploration and development costs. *This means*

Gas," in *Natural Gas Policy Issues and Options*, Senate Interior and Insular Affairs Committee, Serial No. 93-20 (92-55), 1973, pp. 155-157.

[2] Ibid., p. 152.

that the price increases must precede the cost increase. It follows, therefore, that the use of test year costs to justify the necessary price to elicit the required additional supply is destined to be self-defeating. [Emphasis added.]

Regulation Based on "Costs" Will Create Massive Shortages in the Future. Those advocating cost-based regulation, even under the guise of "regulatory reform with other standards added," are effectively guaranteeing that the United States will grow more dependent on imported gas and oil. By 1980, as a result of cost-based regulation, the gap between the quantity supplied and the quantity demanded is predicted to be 59 percent of current consumption levels:

> The "Stevenson Bill" (Senate Commerce Committee bill S-2506) calls for an expansion of regulatory jurisdiction for the Federal Power Commission to cover all of the sales of gas at the wellhead (including intrastate sales). The bill specifically sets out standards for price ceilings based upon historical average costs, so that it is assumed that the goal of this legislation is to control fuel prices to stop the price increases now occurring under more relaxed Federal Power Commission regulation in the last three years. *We infer from the motives of the Commerce Committee that price increases would be limited to approximately 1 cent per year in the future, where justified by changes in average cost of drilling and production that year or some previous year. Under these conditions, we predict that excess demand will be as large as 10 trillion cubic feet by 1980. . . . Under these conditions, the demand for LNG would be quite extensive . . . the amounts involved exceed more than 3 billion dollars per annum, of which it would appear that more than one-half would constitute rate base for purposes of calculating profits of the pipelines. . . .* [Emphasis added.] [3]

Regulated cost-based prices cannot represent competitive market prices, whether set by rule-making procedures or by the lengthy hearing process. As explained by Frank C. Allen:

> Dr. Wilson . . . states that it is absurd to contend that cost-based prices and competitive market prices are somehow totally different economic concepts. His statement may be true within the realm of pure economic theory where producers would be able to produce all the gas that is desired at the lowest possible cost plus a fair rate of return. However,

[3] MacAvoy and Pindyck, statement before the Senate Interior and Insular Affairs Committee, pp. 4-5.

once regulation is introduced into the picture, several factors begin to influence the costs that producers can expect to receive. First, once a producer commits his gas to the interstate market he cannot withdraw without abandonment authorization from the Federal Power Commission. Second, once a rate is established, a moratorium period of several years has in the past been imposed during which time no increased costs can be recouped. Third, price adjustments can only be effected with the approval of the Federal Power Commission, often after long periods of delay. Fourth, in order to justify price increases, proprietary data must often be made public and subjected to scrutiny during protracted public hearings. . . .

Add to these inhibitions to full cost recovery the fact that historical costs without adjustment were traditionally used to establish prospective rates, and it is hard to conceive how such "cost-based prices" can in any sense be considered competitive market prices. [Emphasis added.] [4]

In conclusion, regulation with the "cost" standard is inimical to the goal of reducing energy dependence within the next decade. In fact, a continuation of such regulation will greatly increase the nation's oil and gas imports. Price ceilings based on "cost" are by definition backward-looking, and the arbitrary ceilings chosen will not represent the equilibrium levels required to balance supply and demand. Even if such regulation is applied only to the top twenty producers, it may be pointed out that these producers currently account for about 80 percent of the supply produced for consumers today. Ceilings established on the basis of "cost" will continue to limit their undertakings solely to low-cost projects. As the resulting gas shortage worsens, the demand for oil and coal and pressure on oil and coal prices will increase. Where oil and coal cannot be substituted for curtailed natural gas, unemployment and reduced product output will result.

Regulation: "Economic" Standards

One congressional option is the use of economic factors as opposed to "cost" standards in setting field prices. Since the use of the "cost" standard (either alone or in conjunction with noncost standards) guarantees shortages, the following criteria have been proposed as an option for determining field price ceilings for gas sold interstate:

---

[4] Allen, comments on "Policy Issues and Options Affecting Natural Gas," p. 185.

(a) The current and projected price of other fuels at the point of utilization, adjusted to reflect a comparable heating value;
(b) The premium nature of natural gas and its environmental superiority over many other fuels;
(c) The current and projected prices for the importation of liquefied natural gas and the manufacture of synthetic gaseous fuels; and
(d) The adequacy of these prices to provide necessary incentive for exploration and production of domestic reserves of natural gas and the efficient end-use of such supplies.[5]

There are, however, disadvantages in using economic standard regulation as a solution to the natural gas shortage. First, there is the problem of setting the "appropriate" rate and the equilibrium price required to balance supply and demand. Only by luck could any agency set the "right" price ceiling, regardless of good intentions and more appropriate standards. Competitive prices in the field are set by a variety of market factors affecting both buyers and sellers. While econometric modeling may be helpful, geological uncertainty as well as other factors requiring assumptions and guesswork guarantees that there is no substitute for the competitive forces at work among pipelines and producers in the field. Regulators' estimates of proper price levels are only estimates; competitive field markets set the right price levels, according to the circumstances in each sale, to balance supply and demand.

The use of the national or area price ceiling method in and of itself leads to resource misallocation, moreover, regardless of the level of price ceilings set. This misallocation results from the failure to recognize that different sales involving differing factors call for different price levels.

> The value of a particular reserve of gas to a particular pipeline within a region is a function of its location vis-à-vis a pipeline (the farther away, the greater the gathering cost to the pipeline), its pressure, its size (since large deposits are less costly to process, contract, and to gather and transport), its rate of allowed variability of depletion, and the impurities and British thermal unit (Btu) content of the gas entrapped. The value of these characteristics can be different for different potential pipeline purchasers according to the conditions exhibited by those pipelines. In a free market, each of these reserves presumably could command a price equal to its

[5] Senate Bill S. 3040, "Natural Gas Act Amendment of 1973."

value to the various pipelines, and in the bargaining process contracts would be fashioned that would match reserves and pipelines in an optimal fashion. In consequence each reserve would presumably be priced somewhat differently and not uniformly.[6]

In the unregulated market, such differentials are reflected in the various prices set. Recognizing this, the Federal Power Commission first attempted to set price levels on an individual basis: the practice resulted in an administrative nightmare and was abandoned by 1960.

A second major disadvantage of continued regulation is that there are always unnecessary delays. Three kinds of serious delay would be inherent in any regulatory scheme devised. Administrative delay would continue because those affected by rate changes will demand the rights of constitutional due process and the opportunity to be heard on the question of what they believe to be proper rate levels. Even with new standards, cases would likely take eight to twelve months to resolve. Differences of opinion as to the appropriate rate level also impede action by the regulators.

Third, and most critical, is judicial delay. Past experience suggests that proponents of low wellhead prices would continue their efforts to overturn any price ceiling they deemed "too high." Shifting the burden of proof to the litigant might be helpful, but it would not effectively deter litigation. Opponents of higher prices will litigate to delay price increases, to affect agency attitudes and to deter requests for price increases.

All in all, regulation by "economic standards" falls far short of the mark. More decisive action is urgently needed to reverse the long-term trends that created and are escalating the nation's natural gas shortage. Moreover, if such regulation were applied only to the interstate market, intrastate consumers would continue to be the chief beneficiaries. If regulation were applied to the intrastate market as well, the nation's current energy problems would be severely compounded. Field price regulation for gas sold interstate has already done enough damage. Low field prices for gas sold interstate have historically depressed prices for intrastate gas and other fuels until the advent of severe shortages, thereby indirectly contributing to both the fall-off in domestic energy production and the high demand for energy in the United States.

Since regulation itself inhibits supplies of natural gas, moreover, extending regulation to the intrastate market would amount to amputating the healthy left hand of a patient whose diseased right hand

[6] Milton Russell, "Producer Regulation for the 1970s," p. 229.

has already been amputated so that both sides will match. Continued field price regulation with extension of controls to the intrastate market could only be suggested for political reasons. If representatives of consumer states favor continued field price regulation for political reasons, it would seem that the consumers they represent should bear the burden of the decisions made by their elected officials. It is noteworthy that regardless of their awareness that new gas deregulation will deter business investment in their states, producing-state representatives have joined knowledgeable consuming-state representatives in favoring new gas deregulation.

The Consumer Compromise

The fatal drawback of market simulation regulation is that it cannot simulate the competitive free market: it moves much more slowly than the market, it introduces additional uncertainties, it misallocates resources, and it is far more costly than the free market. Uncertainties, delays, costs, and misallocations of resources could be eliminated if the regulation of the wellhead price of natural gas were eliminated. The way to obtain the results desired from free market simulation regulation is to allow the free market to operate. It can simulate itself far more perfectly than the Federal Power Commission or any other federal agency can simulate it.

Despite the acknowledged superiority of a market solution to the regulatory problems of pricing gas at the wellhead, the political process has made decision making in this area difficult. Political positions have been taken which are difficult for the politicians to abandon without losing "face."

The first error was made by the administration in suggesting a two-tier price system with continued regulation of "old" gas. There is considerable support from the experts that all gas—new and old—should be deregulated. Milton Russell stated that

> the final source of regulation-induced misallocation of resources is associated with the multi-price system. Two allocational losses arise from the multi-price system. First, in this era of rising prices the average price of gas to the consumer is lowered below the marginal cost of its production; hence "too much" is consumed in the present. Second, the consumer whose pipeline receives proportionately more of its gas from an old gas area will consume too much, while the consumer from a predominantly new gas pipeline will consume relatively too little. . . .

. . . effective allocation of resources in an economy depends on the consumers each paying for a commodity approximately its marginal social cost. To the extent it is required by regulation and not by free contract, the differential pricing of gas not only misallocates resources over time and among consumers, but it also transfers income with no obvious policy justification. While on the surface there may be an appealing equity argument that producers who have obtained gas reserves at a low cost should not be allowed to profit from the social appreciation of reserve value, there is a similar argument that particular groups of consumers should not obtain consumer surplus on the basis of the time at which they began to purchase gas. The same issue has even more force when applied to associated gas because here it is strictly fortuity that dictates which consumers will be benefited. Not even contract time is a factor. If the income distribution issue is thus a standoff, then allocational efficiency becomes more appealing as the goal to be sought, and a one-price system is allocationally superior to the multi-price scheme.[7]

In other words, if income redistribution is still desired, taxation rather than continued differential price regulation is the appropriate method to minimize resource misallocation. Moreover, those who argue that higher prices for "old" gas result in "windfall profits at the expense of consumers" fail to take account of the functions performed by prices and ignore energy economics and the facts.

The idea that costs for gas already committed to contract do not increase is false. Gas is customarily contracted for after a field has been discovered and several wells drilled to define the general limits of the field, but before the field is fully developed. Development may proceed over a period of years even after the gas is committed to contract. The drilling of additional wells, including wells to improve the deliverability of the field, proceeds at the costs obtaining in subsequent years, not the costs obtaining at the time the contract was signed. With unexpected inflation, the incentive to fully develop a field at fixed prices declines.

In short, costs continue to rise. Costs rise particularly for the older wells. Work-overs, additional compression and fracturing are often needed just to keep an old well going. Indirect costs (salaries, cost of capital, overhead and equipment costs) also rise. If all these costs cannot be met with increases in prices for "old" gas over the years, wells will be prematurely abandoned as uneconomic, and

[7] Ibid., pp. 230, 233-234.

consumers will lose the gas that could have been produced with higher prices. And time is lost too. The quickest way to increase gas supplies is to increase prices for gas from old fields and old wells. Adding compression to an old well takes far less time than exploring and developing a new area.

Moreover, the notion that any price level above FPC cost determinations leads to a "windfall" is also false. The evidence does not support the view that producer costs have at any time been accurately calculated on an area-wide or industry-wide basis. The facts that exploratory drilling continually declined throughout the 1960s and that the number of independent producers fell off drastically during the same period suggest precisely the opposite conclusion. In nearly all instances, moreover, the FPC chose the lowest of many judgmental alternatives in cost allocations.

Another false view that needs to be refuted is that additional revenues above costs are paid out as dividends to the stockholders of the companies. With some exceptions, additional income deriving from these additional revenues is used as the primary source of capital for additional exploration, particularly for the independents. Joint Association Survey data show, for example, that during the period from 1962 to 1971, producers invested on average 75 percent of their domestic revenues in further domestic exploration, development and production of hydrocarbons. Total revenues from domestic oil and gas operations for 1971 were $13,804,000,000, of which approximately 65 percent or $8,915,000,000 were reinvested for domestic oil and gas exploration, development and production. In 1972, comparable figures were $13,833,000,000 and $10,677,000,000 (77 percent). A large portion of what might be termed "excess profits" is also collected by the U.S. government in the form of taxes.

An important feature of the "windfall profits" myth is the phrase "at the expense of the consumers." Generally "consumers" is a term used to describe residential gas users, although industry and electric utilities use more gas than any other consumer group. When gas sales are curtailed for lack of supply, the industrial consumers are generally the first users to have their supplies cut off. The remaining customers must then pay transmission costs that used to be paid by the industrial consumers. Had higher prices been paid to producers, consumers would get more gas for their money instead of a larger share of fixed costs. The residential consumer also loses if high cost supplemental supplies are added to the gas stream to prevent loss of industrial customers. These higher prices for a less secure supply source add more to cost and prices in the long run than would deregulated prices for natural gas in the field.

In short, with ample analytical support for deregulating all gas, ending price controls on new gas should not pose the political problems it has. This latter alternative is supported by knowledgeable consumers and can be shown to be in the national interest in combatting foreign energy dependence, inflation and recession.

The words of actual consumers are distantly heard, if at all; but on 31 May 1974, Consumers Gas Company, a natural gas distribution facility serving twelve towns in Southeastern Illinois, wrote to its congressman:

> Our customers aren't nearly as concerned with the price of natural gas as they are with the availability of it. Alternate fuels cost them approximately twice the price they are presently paying for natural gas. We need your help in bringing this fact forcefully home to the Federal Power Commission. This must be done by whatever means it takes to accomplish it. Your efforts in support of free enterprise in the energy industries are well founded and appreciated.

The company complained about the "attempt of" the Federal Power Commission "to regulate domestic gas purchase contracts to a price of approximately 40¢ per Mcf." Hundreds of small and large industrial consumers have called for new natural gas field price deregulation.[8]

American consumers will probably pay less for energy if new gas controls are lifted than if they are maintained. The Final Environmental Impact Statement on Deregulation, issued by the Department of the Interior, concluded that continued FPC regulation "could actually lead to higher gas prices to the consumer than deregulation." The Federal Power Commission's views are similar as the following excerpt from a 3 October 1973 letter from FPC Chairman John N. Nassikas to Senator James L. Buckley reveals: "We believe that deregulation of the price of natural gas at the wellhead will result in lower priced energy to American consumers than will result from the present regulatory structure."

New gas decontrol is less inflationary than continuing regulation. Consumers are now facing shortages and can expect additional curtailments—especially large industrial users and electric power plants. These shortages are forcing upon such consumers the choice of either reducing production or, where possible, purchasing alternate fuels. Reduced production forces layoffs and increases unemployment, a

[8] For example, all respondents in a poll of industrial customers conducted by the Columbia Gas Distribution Companies (99 N. Front Street, Columbus, Ohio) favored deregulation.

situation that can have significant recessionary effects on local economies and on national output. At the same time, reduced output creates higher prices, because of the bottlenecks in the short run and then ripple effects throughout the economy.

If a company whose gas has been curtailed can purchase and utilize alternate fuels, however, the cost of production rises immediately. For example, as of May 1974, the average price of fuel to steam-electric power plants, the primary consumers of natural gas, was (in cents per million Btu): gas 44.0, coal 65.8, and oil 187.9. The cost of the final product will tend to increase and, because of the ripple effect, other goods will also be more costly to consumers.

On the other hand, if price controls are lifted on new gas, interstate pipelines will most likely be able to purchase immediately new onshore gas that would have gone to intrastate pipelines. Most of the gas in the pipeline will, for years, be "old" gas, however, at significantly lower prices because of the long lead time (three to five years) in bringing new gas on stream. In addition, the new gas cost will be averaged in with the old gas cost, again lowering the potential inflationary impact.

Over the long term, decontrolling new gas will continue to stimulate supply with a gradual price impact. The increased supply that will result, particularly from the federal domain, will tend to lower natural gas prices. The increased supply will also tend to reduce the need for industrial production cutbacks, and attendant unemployment, as well as the need for substituting costly alternate fuels which impact on the economy immediately. Another benefit would be reduced demand for alternate fuels, which will tend to reduce the price pressure on oil and coal.

Finally, it should be noted that the consumer cost of deregulating new gas, as estimated by David Schwartz in his 25 September 1974 letter to Senator Warren G. Magnuson, was, as Chairman Nassikas put it in the previously mentioned letter to Senator Buckley, "grossly exaggerated and misleading." Chairman Nassikas enclosed with his letter a ten page analysis of the more prominent errors in the Schwartz findings. The concluding section is particularly worth noting:

> Whatever the cost of deregulation may ultimately be it will be far less than the costs associated with current and anticipated curtailments of gas deliveries, idle pipeline capacity, high-cost supplemental supplies, and continued dependence upon foreign sources of supply. Deregulation of new gas, as proposed by the administration will allow market factors to determine price levels and allocate natural gas. Deregula-

tion of new gas committed to the interstate market will more effectively allocate our natural gas resources by:

(1) Eliminating the disparity between the unregulated intrastate price of natural gas comprising one-third of gas consumption and the regulated interstate price comprising two-thirds of gas consumption.

(2) Avoiding distortions in the price of current natural gas supply from–

  (a) pipeline imports ranging to double the regulated price of domestic gas to the same West Coast and Mid-West markets;

  (b) LNG imports delivered to East Coast markets at two-three times the delivered price of natural gas from domestic sources to the same city gate;

  (c) reformed gas and coal gasification at two-three times the price of gas to the same pipeline delivery points.

(3) Equilibrating natural gas supply and demand by an increase in price to the market clearing level as the result of the operation of workably competitive market forces, restraining artificial demand induced by prices established below real market value and stimulating supply by market price incentives for greater investment in exploration, development and production of natural gas.

We would emphasize again that Dr. Schwartz did not attempt to measure the unfulfilled natural gas demand that has been referred to higher priced fuels. For example, current firm natural gas curtailments stand at a level of 1.8 Tcf annually and are expected to continue to rise dramatically in the foreseeable future. This energy, if replaced by oil imports, would be equivalent to about 300 million barrels of oil at a cost of $11 per barrel, or about $3.3 billion. Under deregulation the increased supplies of natural gas needed to fulfill demand will be delivered at prices below those of all other fuels. If this substitution took place even at $1.00 per Mcf, the difference would be a net saving of $1.5 billion.

The curtailments referred to above which amounted to a little better than 10% from April 1973–March 1974 are of firm contractual commitments for interstate pipelines, reflecting very little actual decrease in sales volume. In other words, the curtailments may be attributed, to a substantial degree, to growth in the requirements for natural gas under existing contracts rather than a large constriction in supply. Reserves additions reported in the last six years indicate there will be a substantial, wide-spread constriction of supply so that the existing level of sales can not be maintained in any year after 1973–74, absent an immediate and substantial reversal in the recent trend of supply additions.

This Nation can ill afford a continued exchange of rhetoric, accusations, innuendoes, and recriminations. If we are to maintain our economic stability, we must take measures to bring forth additional gas supplies at the earliest possible time. In my judgment, when all is said and done, we are courting disaster when we continue to impose price regulation on the producing segment of the natural gas industry. . . .

It seems obvious that it is in the consumer's interest and in the national interest to deregulate new gas field prices now.

6
SUMMARY

Nature of the Natural Gas Shortage

As the evidence given in preceding chapters indicates, the natural gas shortage is genuine. Real prices for new natural gas sold interstate fell throughout the 1960s. Discoveries of natural gas peaked at approximately the same time as new contract prices did, and fell with the price decline thereafter. The proportion of natural gas prospects that appeared profitable to explore and develop was progressively reduced. The physical shortage today is primarily the result of problems of deliverability in old wells.

Since about half the natural gas yet to be found and developed is onshore, the shortage is more related to price than to offshore leasing. Price ceilings limit the prospects that will be developed to those that are expected to be sufficiently low in cost to be profitable under the price ceiling. Over time, the cost of finding and developing reserves becomes more expensive, as the least expensive reserves are developed first and the number of sites where low cost discoveries may be expected dwindles.

The Role of Regulation

The natural gas shortage is the inevitable result of FPC field price regulation.[1] Among the problems of cost-based regulation for natural gas are the following:

[1] The list of experts agreeing on this point includes Stephen A. Breyer (Harvard Law School); Thomas Stauffer (Harvard University); Keith C. Brown (Purdue University); James Cox, Arthur Wright (University of Massachusetts); Edward E. Erickson (North Carolina State University); Edmund Kitch (University of Chi-

- Price ceilings based on "cost" reflect the fact that, as far as possible, only "low cost" wells are drilled because of the anticipated low price ceilings. Thus, a cost-price circularity is built into the cost-based regulatory system.
- The range of cost estimates presented in FPC proceedings is wide. FPC choices among them do not represent "in fact" costs. The "cost" figure is always arbitrary.
- Price ceilings based on FPC estimates of average cost make it reasonable to assume that roughly half the gas currently being produced costs more to develop than producers are allowed to receive for it.
- Price ceilings for what is determined (by date of contract of well completion date) to be "new" gas are derived from the same body of data that was used to formulate prior price ceilings. Prices are always set on the basis of historical data.

Economic theory thus supports the data indicating that the natural gas reserve and production shortage not only exists but is the inexorable outcome of field price regulation.

The real wellhead price of natural gas fell 15 percent in the first nine years of field price regulation. Discoveries of natural gas peaked at approximately the same time as did real new contract prices. The current shortage is a direct result of the reduction in the rate of new natural gas discoveries during the 1960s, and this low level of natural gas exploration and discovery was due to FPC ceilings.

The decline of real prices for gas sold interstate indirectly depressed prices for gas sold intrastate and those for other fuels, while accelerating demand until the inevitable shortages became pronounced. The price of natural gas to the residential consumer increased only 19 percent between 1960 and 1972, while, during the same period, the consumer price index increased 41 percent. Price increases in natural gas have also lagged behind price increases for all other fuels and increases in the wholesale price index. Natural gas prices have been so low in relation to other fuel prices that gas demand grew at a higher rate than the rate of growth for total energy. Demand for natural gas has also grown because of the rising prices of other fuels, especially low sulfur fuels. Where gas is in short supply, moreover, the shortage tends to drive up prices of alternate

cago); Clark A. Hawkins (University of Arizona); M. A. Adelman, Paul W. MacAvoy, Robert S. Pindyck (Massachusetts Institute of Technology); Milton Russell (Southern Illinois University); and Robert M. Spann (Virginia Polytechnic Institute).

fuels, such as coal, as potential buyers of gas are forced to increase their buying of these alternative fuels.

The Present

From November 1974 through March 1975, firm natural gas curtailments affecting every region of the nation will increase 80.9 percent over the previous winter to 0.768 trillion cubic feet. Current firm natural gas curtailments are at a level of 1.8 Tcf annually, and the fall-off in reserve additions in the contiguous states in the past six years plus the fall-off in deliverability of old wells indicate a substantial, widespread and rapidly increasing constriction of available supply.

Without an immediate and substantial reversal in recent trends of supply additions, existing sales levels will not be maintained. Even with a moderate growth rate and success in gas additions through supplemental sources, it is estimated that reserve additions must average 30 Tcf (or one and one-half times the 1956 record level for reserve additions in the contiguous states of 24.7 Tcf) each year from 1974–1990.

The downward trend in gas well drilling activity was reversed in 1972. Since 1971, the number of gas wells completed rose from 3,830 to 4,928 in 1972 to 6,385 in 1973; 1,800 gas wells were completed in the first quarter of 1974 as opposed to 1,392 in the first quarter of 1973. Higher intrastate prices and the anticipation of deregulation for new gas sold interstate has undoubtedly effected the upturn in natural gas drilling in recent years. Whereas the number of gas wells completed in 1972 is the all-time record, it is noteworthy that the percentage of gas finds that after six years prove to be of significant size has generally declined from a high of 52 percent in 1945 to 23 percent in 1965. A reversal in supply additions will require massive investment and, because of the three-to-five year lead time, policy changes *this year* will be required to avoid the extremely serious supply disruptions predicted for the future.

This year the cost to consumers of congressional inaction on natural gas legislation amounts to $3.3 billion (1.8 Tcf is equivalent to 300 million barrels of oil which, at $11 per barrel totals about $3.3 billion). Even if all deregulated new gas prices rose to $1.00 per Mcf, the net saving to energy consumers and the nation would be $1.5 billion.[2]

[2] Attachment to the 3 October 1974 letter from FPC Chairman John N. Nassikas to Senator James L. Buckley.

The direct impact of the natural gas shortage where alternate fuels cannot be substituted is unemployment and reduced national output. Supply disruptions in Pennsylvania in October 1974 caused the temporary lay-off of 600 employees and, had it not been for emergency relief, the lay-off was predicted to rise to 6,000 in a matter of a few weeks. The state of New Jersey projects a lay-off of 120,000, directly as a result of the gas shortage if curtailments rise from 28 percent to 30 percent. In the Baltimore area, a recent curtailment order would have resulted in the loss of 10,000 jobs according to a recent study. Where oil or coal can be substituted, it will be at high cost to consumers. But in many instances natural gas consumers do not have the capacity or storage facilities required to use other fuels.

The Future

Field price regulation based on cost with the addition of a few cents per Mcf for "noncost" factors guarantees far more severe natural gas shortages in the future.

Rationing or "allocation" is not a solution in that it merely redistributes shortages, besides being unworkable from a practical standpoint for a variety of reasons:

- Natural gas contracts are usually made for twenty years or more. Litigation against the federal government would fill the courts if abrogation of existing contracts were forced on pipelines and consumers.
- There are physical problems that make allocation of intrastate gas to the interstate market not a practical solution. Natural gas cannot be transferred from one line to another where there are no physical connections, nor can it be moved about in trucks, except in the rare circumstance where it is liquefied (at high cost to the consumer).
  —In the field, the only pipeline that reaches the well is the pipeline connected to the well belonging to the company that bought the gas.
  —In transmission, natural gas flows through a complex web of pipelines, and many companies are involved.
- If it were physically possible to take gas away from consumers in Louisiana, for example, the consequences might well penalize that region: unemployment where factories do not have alternate fuel capability, reduced electrical output, reduced production, higher costs and increased air pollution.

- Allocating natural gas supplies would penalize those who had planned well for the future, both consumers and gas companies. Such an allocation program would also reward those who, through poor planning, had allowed their supply situation to deteriorate.
- Allocating gas supplies simply postpones the day of reckoning. We need more supplies to distribute, not more shortages to allocate.
- An allocation program would substantially weaken the incentive to be resourceful in finding and developing new sources of gas. Robbing Peter to pay Paul means that Peter will no longer seek his own supplies. Then Peter *and* Paul—and the consumer—will be without supplies.

On the basis of the extensive studies that have been made, it has been concluded that field price regulation is not needed and has hurt consumers more than it has helped them.

- Field markets for natural gas are workably competitive.
- The primary beneficiaries of field price regulation have not been residential consumers, but industrial consumers in the producing states where field prices are not regulated.
- Regulation is an inefficient allocator of resources.
- The natural gas shortage is the inevitable result of field price regulation.

Continued regulation—even without the so-called cost standard—cannot adequately "solve" the gas shortage. Besides low prices, the keynote of the regulatory scene has been uncertainty. This uncertainty deters investment because the risks and costs involved in finding significant volumes of gas have increased, as has inflation. Under any option involving continued regulation, there will be continued uncertainty. Even with the use of rule-making techniques, the FPC continues to take about eight months to decide one case. Because of long-established regulatory procedures, practices, and history, the FPC is ill-equipped to act quickly and decisively, whether or not rule-making techniques are used. Years of litigation over FPC decisions will also continue to deter investment.

Most economists therefore do not see more regulation (with or without extension of jurisdiction to the intrastate market) as a viable solution. Sustained regulation of the interstate market (without extension of jurisdiction) will continue to allow intrastate purchasers to buy virtually all of the new onshore gas at the expense of interstate consumers. Yet, with regulation the direct cause of the natural gas shortage, extension of federal field price regulation to intrastate sales can hardly be the answer either. It would result in: reduced investment in natural gas exploration, development, and production; further reductions in supply; deeper curtailments; further industry cutbacks; further unemployment; further dependence on higher-cost imported energy; reduced national output; and further resource misallocation.

If the United States is serious about the need to increase energy self-sufficiency, new gas field price deregulation will be the first step. The enactment of legislation to free controls on new gas prices will result in lower cost energy for all consumers. It will also benefit consumers and the nation in combatting recession resulting from production cutbacks and unemployment where alternate fuels cannot be substituted for dwindling natural gas supplies. It is also less inflationary than turning to imported oil to fill the gas gap. It will cut the price pressure on alternate fuels caused by the deepening gas shortage, as well as spur the development of needed energy resources for the future.

To knowledgeable people, the choice is clear, and too much time—twenty years—has already been wasted in the regulatory experiment. It has failed to meet consumers' needs and national needs. The time has come to close debate and to act decisively. While total deregulation is preferable to new gas deregulation, half a loaf is better than no bread at all.

Cover and book design: Pat Taylor